SCIENTIFIC ATTITUDE AND HIGHER EDUCATION

By

J. Vidya Sagar

ANMOL PUBLICATIONS PVT. LTD.
NEW DELHI - 110 002 (INDIA)

ANMOL PUBLICATIONS PVT. LTD.
4374/4B, Ansari Road, Daryaganj
New Delhi - 110 002
Ph.: 23261597, 23278000
Visit us at: www.anmolpublications.com

Scientific Attitude and Higher Education

First Edition, 2003

ISBN 81-261-1560-2

PRINTED IN INDIA

Published by J.L. Kumar for Anmol Publications Pvt. Ltd., New Delhi - 110 002 and Printed at Tarun Offset Press, Delhi.

Contents

Preface

Modern age is rightly termed as Scientific era. At no point of time in the history of mankind did science dominate human life so overwhelmingly and so intricately as it does today. It has entered all aspects of human life. Technology, which is the application of science has transformed human life and has reduced this huge globe into a small village bringing man closer to man. Industrialisation, by exploiting the nature's potentialities, has brought instant prosperity. The fields of medicine have been revolutionised. Epedemics have been almost controlled. Life span of man is ever lengthening and material comforts have been multiplying. Revolutionary changes have been brought about in the field of communicational technology. Human engineering is a new dimension which has tremendous potentialities of transforming the quality of human substance.

All these have been made possible because of the great strides man has made in the field of Science. Yet Science has tremendous potentialities of making such miraculous advances which we may not be in a position to conceive now.

But the scientific growth is both preceded and followed more vigorously by formation of appropriate attitudes. Such attitudes are termed as Scientific attitudes. It is even called Scientific temper. It signifies a frame of mind and attitudinal structure which are conducive to vigorous scientific research and discovery. The scientific temper and the growth of science are interdependent. The growth in one sector often leads to growth in another sector.

While we make systematic effort for the dissemination of scientific knowledge and testing the achievement of the same in the clientele, we do not care to assess the commensurate growth of scientific temper which is often thwarted by certain extraneous factors like faith, tradition, mythology, religion etc., which do not necessarily have rational bases. So it is no wonder that there is no one-to-one correspondence in growth in science and growth in Scientific Attitudes which ultimately hinders the smoothness of the growth of science. Kepler's obsession with divine music emanating from the drifting stars and Newton's acceptance of transcendental theology without questioning it are but two examples of the unscientific attitudes of great scientists. Each scientist suffers from a hangover of the tradition which invariably hinders the growth of science to some extent. And so to facilitate the accelerated growth of Science there is a need to monitor the corresponding growth of Scientific temper.

Attempts, made to measure the scientific attitude have been relatively recent. All such efforts were made in recent decades in the West. But not many scales have been constructed even in the West to assess the development of Scientific Attitudes at various levels. Still fewer attempts have been made in India in this direction.

The Universities and research organisations are the places where high level scientific knowledge is disseminated and concentrated efforts made towards the scientific discovery. In the interest of growth in science, assessment of the scientific knowledge and the corresponding development of scientific attitudes at higher levels should be undertaken at frequent intervals. The absence of appropriate tools to measure the Scientific Attitudes at the higher level has provoked the construction of the Scientific Attitude Scale by the present researcher as a part of his Doctoral Research. The construction and standardisation of the Scientific Attitude Scale is reported in the following pages.

The report presented also includes the administration of the standardised scale on the Academics and the Research Scholars working in certain prestigious higher educational institutions in Telangana. The influence of vital factors like the substance of each academic discipline, research background, professional and non-professional background, political commitment etc. on Scientific Attitude was assessed. The findings of the investigation thus carried out are presented in the report.

I thank Prof. S. Parmaji, former Dean, Faculty of Education, Kakatiya University, for guiding the research. His association has not only been inspiring but so sustained that he must share in equal measure the credit for successful completion of this research. And so he has kindly accepted to lend his name as co-author of this book.

I must place on record my indebtedness to Prof. M.Malla Reddy, the then Dean, Faculty of Education, and the present Vice Chancellor, Osmania University, for providing the opportunity of conducting the research under report at the college of Education, Osmania University. I must thank Sri Sundaresh Peri, Lecturer in Statistics at Kakatiya Medical College, Warangal for the valuable discussion we had about the statistical techniques used and for computerising the data.

I must thank my wife Smt. Latha, and my children Kavitha and Vinay for the cooperation they have extended in completing the present research assignment.

Finally I must thank the Mr. J.L.Kumar New Delhi, for undertaking to publish the research report in the shape of a book.

Hyderabad
12.11-2002.

Jamalpur Vidya Sagar

List of Figures

List of Tables

1
Introduction

Science is concerned with the facts of the World we live in. When the facts of the world are observed or experienced and when they are put in a systematic order they become Science. Thus Science is a systematised body of empirical knowledge.

Science is finally interested in understanding the relationships between facts. It is assumed that both the facts about the phenomenal world and their interrelationships are relatively enduring implying an order and uniformity which can be verified. This order is studied in terms of cause-and-effect relationships. Either the effect could be predicted or the cause post-dicted, the implication being that the inherent order in the universe is governed by the law of Causality. It is postulated that a phenomenon occurs only because of the presence or absence of a set of antecedent phenomena.

Thus Science is interested in the universe of facts and the universe of their relationships. The relationships are inherent in the combination of facts, and the facts and their interrelationships are complementary to each other. The relationships between facts, both physical and social, are called Laws. When related laws are put together they form a theory.

The knowledge of facts, laws and theories which underscore an understanding of the world from the body of Science. The greater the quantum of Science, the greater the understanding of the world, which again implies an increased facility to predict the events and to control the consequent phenomenon by manipulating the antecedent phenomena. The ever increasing scientific knowledge, including the knowledge of facts, laws and theories, with their predictive and post-dictive implications, has transformed human life. Particularly, the application of science, which is termed as Technology, has transformed human life beyond recognition.

Thus Prometheus unbound, man's quest has secured to him immense access to the secrets of the nature, and with the help of Technology, the applied form of Science, he has performed many a conquests taming and harnessing the potentialities of nature in the service of man. Man's life has changed qualitatively ever since his quest started comprehending the nature and its working.

From hunting to domesticating the animals, from fruit gathering to the pursuit of agriculture, from using stones to the use of metals, there was a progress, though sluggish in the beginning. The progress got accelerated with the invention of the steam engine. The genie hidden in the thermo-nuclear potentialities of the nature is being commissioned incessantly and in a mighty way to serve the cause of man not only in enlarging the productive functioning of the soil thereby removing poverty, and hunger from the world but in providing physical comforts to an ordinary man in such a way that the mightiest emperor of the past might feel jealous. Man's quests and conquests have shrunk the world into a global village. He could control the deadliest diseases because of the developments in his scientific knowledge.

Man has been making great strides in a variety of fields in multiple ways because of the expanding horizons of his scientific knowledge. He can today dam the mightiest rivers and channel their humbled pride to irrigate the distant parched lands. Overnight he can get the more cosy and comfortable homes erected than Alladdin's genie could have done. Not only can he evolve hybrid seeds which can ensure doubly redoubled production, he can commission a test tube baby to the desired specifications. He can rocket to distant heavenly bodies which his predecessors could not dream of. The revolution in the printing technology has brought all human knowledge within the reach of all men, who want to have it. The marvels achieved in the field of communication technology has brought the most distant objects within the seeing and hearing range of man.

Simultaneously Science has freed man from the spell of superstition and transcendental ethics. By postulating theories of origins of earth and man, it has liberated him from the clutches of religious mythology. Science is interested in the secular knowledge which is empirical in nature. Thus modern sciences help man in evolving humanistic values which have world-wide ramifications rather than sectarian and small group oriented.

One extra-ordinary feature of man's quest and conquest of the nature is that it gets progressively accelerated with the possibility of the entire human knowledge doubling itself within a space of a few years. Unsatisfied, man wants to further accentuate the acceleration of the quest and conquest of the nature.

But the scientific quest is predicated on certain wide ranging attitudinal antecedents which constitute Scientific Attitude, Scientific Mindedness or Scientific Thinking. The objectives of teaching of Science include not only imparting scientific knowledge but scientific attitudes which are complementary to each other. But though the knowledge of science is consistently tested, not many devices or tests are evolved to test the Scientific Attitude.

Universities are the citadels of higher learning. Most of the teaching of Science takes place in the portals of such citadels. Particularly the Academics and Research Scholars are engaged in the pursuit of research, employing scientific method which should further consolidate the scientific attitudes.

But does it invariably happen? It is said that a Scientist of Newton's calibre suffered from the hangover of the tradition and professed equal interest in transcendental Theology while Kepler heard divine music in the motions of spheres (Shah, 1964).

Thus a scientist is not only exposed to the scientific knowledge but he bears impact of the traditional culture which may not be in tune with the scientific thinking or scientific mind. The content they handle, the conceptual framework it provides and the exclusive clientale the subject area provides to the specialist in the university might colour his perception and world view. Thus the Academics working at higher levels in the universities may display Scientific Attitude in varying measures. If it is so what are the academic and biographical factors which are conducive to the development of Scientific Attitude?

Such studies are not often undertaken. Nor are the appropriate tools available to measure Scientific Attitude of the Academics and Research Scholars at the universities. Thus a need to construct and standardise a tool to measure Scientific Attitude and find out how Scientific Attitude is related to the content background of the Academics and Research Scholars in universities and to their biographic antecedents.

A detailed discussion covering the definitions of Science, Scientific Method and Scientific Attitude are presented in the following paras for a deeper understanding and appreciation of the problem.

What is Science?

Science is a complex term which cannot be explained through one definition. And so, an attempt is made to explain "Science" through the following definitions:

"Science is an accumulation of systematized facts".

Science has inherent urge to know and understand facts and it puts the facts together to arrive at meaningful theories and laws through rational approach based on observation, reflection and reasoning. When new facts are discovered and a new reasoning is developed, the old conclusions are revised and reformulated. Thus, Science comes out with new predictions with the help of newly developed theories and laws. The entire process of Science has been diagrammatically depicted by Ronald D. Anderson et al (1970) which is as follows:

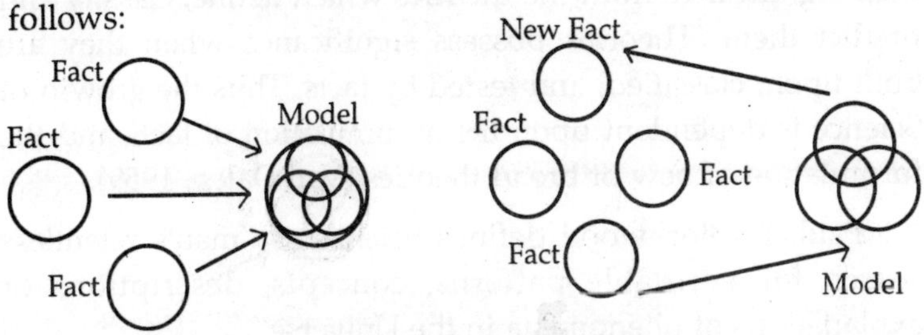

The present investigator has improved upon the said model to make the graphic representation of cyclic evolution of Scientific theory more explicit.

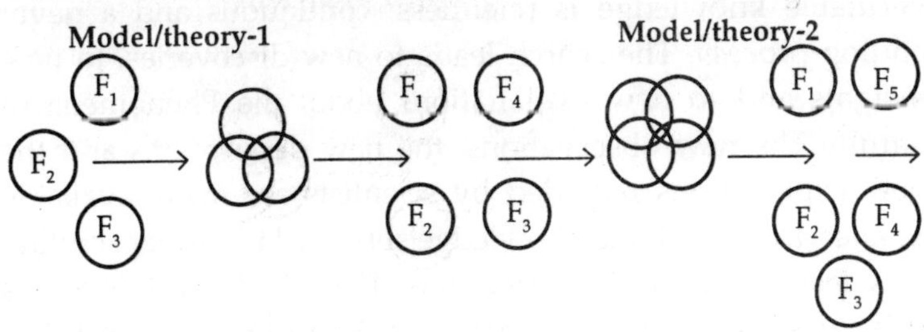

Fig 1. Graphic representation of cyclic evolution of Scientific Theory.

At the initial stage the Scientist is confronted with the facts 1, 2 and 3. The investigation covering these facts leads to the emergence of a model/theory. But subsequent investigators would not only take into account the three erstwhile facts but the newly evolved model/theory as the fourth fact. Besides, a set of few more new facts might be identified (F_5). In the process of new investigation, the investigator will include all the five facts in his investigation and would evolve a new modular theory which again could be treated as new fact (F_6). Thus the Scientific theorisation would progress in a cyclic way adding new facts in its process.

And thus facts and theory are interrelated. Facts without theory or theory without facts lack significance. Facts take their significance from the theories which define, classify and predict them. Theories possess significance when they are built upon, classified, and tested by facts. Thus the growth of Science is dependent upon the accumulation of facts and the formulation of new or broad theories (Van Dalen, 1959).

Paul, E. Blockwood defines Science as "man's relentless search for verifiable patterns, concepts, descriptions or explanations of phenomena in the Universe".

According to him, science is a human enterprise in which men and women always search in their investigations and inquire about verifiable knowledge. The search to seek verifiable knowledge is relentless, continuous and a never ending process. The search leads to new discoveries, to new insights and to new explanations about the Phenomena of nature. The new observations, the new descriptions and the new explanations recorded by scientists are made available as knowledge and this knowledge subjected to further scrutiny may lead to further investigations. Thus the Science is man's relentless search for verifiable knowledge about the phenomenon of nature.

"Science is defined as What Scientists do".

Scientist carries out experiments, observes natural and social phenomena, thinks furiously for long stretches of time to solve a problem, or allows the problem to lie at the back of his mind for weeks together and discovers one day the solution in the form of a new hypothesis or a new method that strikes him like a flash of lightening. At different stages of his work he may be doing any of these things. At one and the same time, too, he may be doing more than one of them. (Shah, 1964).

It is generally said that the scientists perform the following three basic functions.

(i) Scientists make descriptions

Scientist attempts to describe what is, how things are, what things are like, how they change and how they interrelate. Scientist attempts to answer such questions.

(ii) Scientists make explanations

Scientists attempts to tell "Why" certain events and phenomena occur the way they do. This involves a close study of interrelationships. What precedes what? What follows what? Under what conditions do certain phenomena occur? Scientist attempts to answer all these questions.

(iii) Scientists make predictions

Scientist always wants to test the newly found principles in many situations in order to make knowledge more widely applicable and to extend our confidence in its validity. The extension of our knowledge to new situations involves prediction. Making predictions is an important part of what scientists do.

In making descriptions, explanations and predictions scientists use their minds; they use ideas of their own and ideas of others as tools for testing and gaining knowledge. They use many resources to get valid answers to their questions or solutions to their problems. They may invent new tools with which to observe or to check phenomena more accurately. Thus, Scientists do many things in relation to making valid descriptions, explanations and predictions.

According to Columbia Encyclopaeda (1963) "Science is an accumulated and systematized learning in general usage restricted to natural phenomena. The progress of science is marked not only by an accumulation of facts, but by the emergence of scientific method and of the Scientific attitude."

Science, then, is a combination of both "processes" and "products", related to and dependent upon each other. When used in this way, Science offers methods of inquiry useful in learning more about the universe and its working. The processes of Science are Scientific attitudes, and methods of inquiry. As a result of these processes we derive the products of Science which are accumulated and systematized bodies of knowledge. The interrelated and expanding nature of the processes and products of science has rightly been depicted in a figure by Arthur A Casin and Robert B. Sund (1970) in Fig. 2 on next page.

As man uses the scientific processes to investigate the natural phenomenon, he gains increasingly tested Scientific products i.e., facts, concepts, principles, generalisations, theories and laws. These newly acquired products are then used to broaden and increase his scientific investigations which involve further use of processes resulting in more tested Scientific products infinitum.

The Characteristics of Science implied in the definitions stated are summarised as follows:

- Science always studies interrelations between facts.
- Science is self corrective in nature.
- Science is objective in nature.
- Science is based on assumptions.
- Prediction is one of the important features of Science.

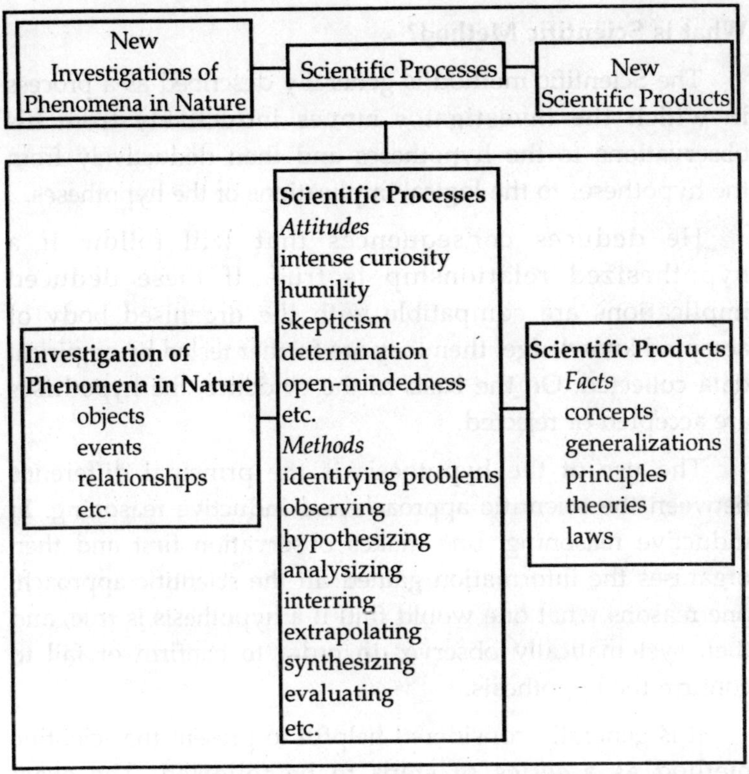

Fig. 2. Interrelationship Between Scientific Processes and Products.

— Explanation of verifiable Knowledge is a basic function of Science.

— Science is both product and process.

— Science is based on the Principle of "Only certainty is uncertainty".

— Science is developed through the Principle of "Positive approach to failure".

— The basic aim of Science is theory building.

— Science is a method of inquiry/discovery.

What is Scientific Method?

The Scientific method is generally described as a process in which the investigator moves inductively from his observations to the hypotheses and then deductively from the hypotheses to the logical implications of the hypotheses.

He deduces consequences that will follow if a hypothesized relationship is true. If these deduced implications are compatible with the organised body of accepted knowledge, then they are further tested by empirical data collected. On the basis of the evidence, the hypotheses are accepted or rejected.

The use of the hypothesis is the principal difference between the scientific approach and inductive reasoning. In inductive reasoning, one makes observation first and then organises the information gained. In the scientific approach, one reasons what one would find if a hypothesis is true, and then systematically observes in order to confirm or fail to confirm the hypothesis.

It is generally considered helpful to present the scientific method as a series of steps to be followed. The exact formulation of the steps will be found to vary from author to author. However, the names given to steps are not important. What is important is to convey the idea that the Scientific method is a systematic process of inquiry involving interdependent parts. It is a method of inquiry that developed gradually over a time and has maintained because it has proved to be a successful method for understanding our complex natural phenomenon.

The five steps in the scientific method are as follows:

1. Definition of the problem.
2. Statement of the Hypothesis.
3. Arriving at implications of the Hypothesis through deductive reasoning.

4. Collection and analysis of the data.
5. Confirming or rejecting the Hypothesis.

1. Definition of the Problem

A Scientific inquiry originates from a problem or a question in need of solution. In order to be a subject of scientific investigation a question must have an essential characteristic. It must be possible to formulate it in such a way that observation or experimentation in the natural world can provide an answer; question involving choice or values can not be answered on the basis of the factual information alone-consider the question; "Do children who are introduced to reading through the Initial Teaching Alphabet (i/t/a) score higher on reading achievement after they have made the transition to conventional letters and spelling than children who are introduced to reading through the Sight method?" This could be investigated empirically by comparing scores on a criterion of reading achievement for two groups who are equivalent except that one group was introduced to reading through i/t/a and the other through sight reading. However, the question "Is training with i/t/a good for students?" Could not be investigated scientifically since it is impossible to know what "good for students" means or how to observe or measure "goodness". Value judgements are out of place in science; so words that imply value judgements should not be included in defining a problem.

2. Statement of a Hypothesis

The next step for the investigator is to formulate a hypothesis, which provides a tentative explanation of the problem. This stage will require a review of the related literature and much further thought. In the example stated the investigator might hypothesize-introducing reading through i/t/a produces greater reading achievement than introducing reading through the sight method".

3. Deductive Reasoning

Next, the investigator deduces the implications of the hypothesis that he has suggested–that is, what would be observed if the hypothesis is true. If it is true that introducing reading through i/t/a produces greater reading achievement than introducing reading through the sight method, then one should observe higher reading achievement scores among those students who were introduced to reading through i/t/a than among equivalent students who were introduced to reading through the sight method.

4. Collection and Analysis of the Data

The hypothesis, or more precisely its deduced implications are tested by collecting data relevant to it through observation, testing and experimentation.

5. Confirming or Rejecting the Hypothesis

Once the data have been collected, the results are analysed in order to determine whether the hypothesis is true or not. In the scientific approach one does not claim to prove a hypothesis; this would be dealing in terms of absolute truth, which is not a characteristic of this approach. One merely concludes that the evidence does or does not support hypothesis.

What is Scientific Attitude?

The Scientific Attitude is adoption of a particular approach to solve problems, to assess ideas and information and to make decisions. Using this particular approach evidence is collected and evaluated objectively so that the idiosyncratic: prejudice of the one making the judgement do not intrude in decision making. No source of relevant information is rejected before it is fully evaluated and all available evidence is carefully weighed before the decision is made. If the evidence is considered to be insufficient then the judgement is suspended until there is enough information to enable a decision to be made. No idea, conclusion, decision or solution is accepted just because a particular person makes a claim but it is treated sceptically and critically until its soundness can be judged according to the weight of evidence which is relevant to it. A person who is willing to follow such a procedure (and who regularly does so) is said to be motivated by the Scientific Attitude.

Some writers lable this attitude as "Scientific mindedness" (Burnett, 1944), "the habit of Scientific Thinking" (Noll, 1933 a) or "the spirit of Science" (Educational Policies Commission, 1966) and it is most often characterised by a list of Component attitudes ("Scientific Attitudes") such as Objectivity, Open-mindedness, Scepticism, and a willingness to Suspend Judgement if there is insufficient evidence. There is general agreement among investigators that a person who has a scientific point of view (1) looks for the natural causes of events; (2) is open-minded towards the work, and opinions of others and towards information related to his problem; (3) bases opinions and conclusions on adequate evidence; (4) evaluates techniques and procedures used and information obtained; and (5) is curious concerning the things he observes. (Fifty-ninth Year Book of the National Society for the Study of Education, 1960).

Heiss (1958) stated that the development of the Scientific Attitude and the ability to use the method of Science are the major goals of Science instruction.

In reporting a research of present nature which envisages the construction and standardisation of a scale to measure Scientific Attitude, needs elaborate treatment. Since the second chapter is exclusively devoted to the reporting of the construction and standardisation of the Scientific Attitude Scale, in the fitness of the research reporting, it was felt appropriate to provide a detailed treatment of the "Scientific

Attitude" as an introduction to the reporting of the construction of the scale.

Problem

The Scientific Attitudes, it is generally said, are the outcome of learning 'Science'. Science thus develops the culture of Scientific Attitude—a culture which is inherent in Science itself. So, a person who has gone across 'Science Course' for some time should certainly resemble in its culture in the form of learning some of the Scientific Attitudes. And so does a person who has gone across through the Scientific Method should also invariably make him to resemble in its culture in the form of learning some of the Scientific Attitudes.

Do persons who have Science background have more Scientific Attitude than persons who do not have Science background? Does Science actually influence the development of Scientific Attitudes? If it influences, is this influence significant? Besides the Scientific knowledge there may be other factors, which influence the Scientific Attitude. These may include demographic factors, and the persons' exposure to a variety of ideologies, discipline, etc.

Thus, the main objectives of the present study were:

(1) to construct and standardise a Scientific Attitude Scale (Construction part), and

(2) to measure Scientific Attitude among the Academics and Research Scholars in Universities in relation to their demographic factors (Testing part).

Significance of the Problem

It has been stated earlier that knowledge of Science and Scientific Attitude are interdependent and complementary. Scientific discovery presupposes a questioning mind and a specific approach which is termed as Scientific Method. Scientific Knowledge on the otherhand not only widens the mental horizons of the learner but it makes him further inquisitive and scientific minded.

Since the Scientific Knowledge is expected to grow in a parallel way along with the growth in Scientific Attitude, it may not happen generally. The influence of tradition some times might stunt the growth of Scientific Attitude. Such a lopsided growth may not be conducive to the growth of Scientific Knowledge. Since Scientific thinking and Scientific Attitude are instrumental in expanding Scientific Knowledge, there is a need to frequently assess the relative growth of Scientific Attitude particularly in the practitioners of Science. Such an exercise requires valid and reliable tools for the measurement of Scientific Attitude at various levels including universities. There is a greater need to construct and standardise a Scale to measure Scientific Attitude in the Academics and the Research Scholars particularly at the University level as they are normally charged with the responsibility of pursuing Scientific Research.

So any attempt to standardise a Scientific Attitude Scale for the Academics and the Research Scholars at the University level is not only desirable but would help in giving a needed thrust to the development of Scientific Attitude by identifying the areas of deficiencies and by focussing the need for concentrated effort to wipe out such deficiencies. Hence the significance of the present research.

Besides standardising Scientific Attitude Scale the research envisages to find out the relationships between various demographic factors of the Academics and Research Scholars and the levels of their Scientific Attitude. If any demographic factor is identified as not being conducive to the development of Scientific Attitude in the subjects the School and College can take compensatory measures to overcome such deficiencies.

Statement of the Problem

The research under report has addressed itself to answer the posers listed in the following paras:

1. Generally it is conceived that women are more traditional than men. Then obviously men should score higher on any Scientific Attitude Scale than women.

But women are exposed equally to the rigours of Academic discipline on par with men in a University situation and pursue Scientific Research in the same way as men. Then would they score significantly less on Scientific Attitude Scale? or will there be no significant difference between the male and female Academics and Research Scholars in their scores on Scientific Attitude Scale?

2. In Universities and higher level colleges one comes across two types of functionaries, namely, Academics and Research Scholars. Academics not only teach but they pursue Scientific Research and guide the Researchers who undertake research. Researchers are normally young and inexperienced. They do not generally teach in the class rooms-teaching which adds to their academic maturity. But it is possible that being young they are relatively less tradition bound.

Within these two groups of functionaries who score more on the Scientific Attitude Scale? Do the Academics score more or the Research Scholars?

3. The age of the individuals is likely to play both a favourable and unfavourable role in developing the Scientific Attitude. With age the Academics are likely to get greater academic maturity and gain more experience in Scientific Research. But again the Age is likely to take them towards tradition. It is not established clearly if the age plays a favourable or an unfavourable role in fostering the Scientific Attitude. Hence the question: Does the Age have anything to do with Scientific Attitude?

4. It is not established clearly if Income has anything to do with Scientific Attitude of the Academics and Research Scholars.

The business and the industrial community which earns more than any other community is found to be more traditional. Have the income differences among the Academics anything to do with their scores on the Scientific Attitude Scale? The research sought an answer to this question.

5. The content background of the Academics and Research Scholars is likely to colour their thinking, and influence their vision by providing a different conceptual frame-work, approach, and association with an exclusive outside world. Each discipline likewise is likely to fashion the vision in its own way. In such a context will all the Academics and Research Scholars with different backgrounds score identically on the Scientific Attitude Scale? Or to put it precisely, has the content background of the Academics and Research Scholars anything to do with scores on Scientific Attitude Scale?

6. Like Age, Experience of the Academics and Research Scholars is likely to play a role in fashioning Scientific Attitude. Will the more experienced Academics and Research Scholars be higher scorers on Scientific Attitude Scale? Or will the Experience in combination with Age blunt the teeth of Scientific Attitude? Or has Experience anything to do with the Scientific Attitude?

7. Within the Universities and Colleges two types of Academics and Research Scholars operate. They are Academic degree holders and Professional degree holders. Their Academic and Professional degrees may not exert identical influence on their Scientific Attitude. Besides, their rapport with their exclusive clienteled may influence their Scientific Attitudes. But who would score more on the Scientific Attitude Scale? Will there be any significant differences in the scores of

Academic and Professional degree holders on Scientific Attitude Scale?

8. It can't be ruled out that the some of the Academics and Research Scholars would be committed to various shades of political ideology. Will the political ideology, left, centre, or right, will influence the scores of the subjects on the Scientific Attitude Scale? Or is there any relationship between the ideological background of the Academics and Research Scholars and their scores on Scientific Attitude Scale?

9. Even the ideological commitment would imply the operation of the committed at various levels. They could operate at the level of Theorisation or at the level of Organisation, or at the Cadre level, executing the orders of their higher-ups. Will the operation of the subjects at different levels have an impact on Scientific Attitude scores? Or the level of Political activism anything to do with Scientific Attitude scores?

10. Finally, it is generally believed that social sensitivity makes one committed to political ideology and an insensitive person remains apolitical. Does the political commitment add to Scientific Thinking? Or the ideologically committed Academics and Research Scholars score significantly higher on the Scientific Attitude Scale than the Academics and Research Scholars who are ideologically not committed?

Objectives of the Present Study

Besides the construction and standardisation of a Scientific Attitude Scale, the present study intended to:

(1) find out the relationship between sex and the Scientific Attitude;

(2) find out the relationship between the occupation of the subjects and their Scientific Attitude;

(3) find out the relationship between the age background of the subjects and their Scientific Attitude;

(4) find out the relationship between the income levels of the subjects and their Scientific Attitude;

(5) find out the relationship between the main content background of the subjects and their levels of Scientific Attitude;

(6) find out the relationship between duration of the experience of the subjects and their levels of Scientific Attitude;

(7) find out the relationship between the mean Scientific Attitudes of Academic and Professional degree holders;

(8) find out the relationship between the political ideological background of the subjects and their Scientific Attitude;

(9) find out the relationship between the levels of Political activism of the subjects and their Scientific Attitude; and to

(10) find out the relationship between the mean Scientific Attitudes of the subjects who had background of some political philosophy and of the ones who had no background of any political philosophy.

Hypotheses

To find out the relationships between the pairs of variables detailed, the following null hypotheses were formulated.

(1) There will be no significant difference in the mean Scientific Attitudes of Males and Females.

(2) There will be no significant difference in the mean Scientific Attitudes of Academics and Research Scholars.

(3) There will be no significant difference between the age levels of the subjects and their mean levels of Scientific Attitude.

(4) There will be no significant difference in the mean levels of Scientific Attitude of Academics and Research Scholars because of the differences in their income levels.

(5) There will be no significant difference in the mean levels of Scientific Attitude of Academics and Research Scholars because of the differences in their content background.

(6) There will be no significant difference between the duration of the experience of the subjects and their mean levels of Scientific Attitude.

(7) There will be no significant difference between the mean Scientific Attitudes of Academic and Professional degree holders.

(8) There will be no significant relationship in the mean levels of Scientific Attitude of the subjects because of the differences in their political ideological backgrounds.

(9) There will be no significant difference in the mean Scientific Attitudes of subjects because of their levels of Political activism.

(10) There will be no significant difference between the mean Scientific Attitudes of the subjects who had background of some political philosophy and of those who had no background of any political philosophy.

The first chapter thus not only introduced the topic but has detailed the objectives and the hypotheses. The Second chapter is devoted to the presentation of the process of the standardisation of the Scientific Attitude Scale.

2

Review of Related Literature and Construction of Scientific Attitude Scale

The main objective of the present study was to construct and standardise a Scientific Attitude Scale to measure scientific attitude among the Academics in Universities and Research Scholars of various faculties in relation to various demographic factors. So, it was thought worthwhile to review the related literature pertaining to this field. The review revealed that Curtis (1924) constructed the first instrument for measuring the Scientific Attitude. The test was based on a definition of the Scientific Attitude consisting of four components, namely, (i) conviction of universal cause-and-effect relationships, (ii) Habit of delayed response, (iii) Habit of weighing evidence, and (iv) Open-mindedness.

Noll (1935) constructed an instrument based on another definition consisting of six components of the Scientific Attitude, namely, (1) Accuracy, (2) Intellectual honesty, (3) Open-mindedness, (4) Suspended judgement, (5) Looking for cause and effect relationships, and (6) Criticalness, including self criticism.

A comprehensive definition of Scientific Attitude with behavioural aspect of each dimension which was spelt out by the Forty-sixth Year Book of the National Society for the Study of Education (1947) is as follows:

1. Looks for the natural causes for things that happen

(a) Does not believe in superstitions such as charms of good or bad luck;

(b) Believes that there is no connection necessarily between two events just because they happen at the same time or one after the other.

2. Is Open-minded towards work and opinion of others and information related to the problem

(a) Believes that truth never changes, but ideas of what is true may change as one gains better understanding of that truth;

(b) Revises opinions and conclusions in the light of additional reliable evidence;

(c) Listens to, observes or reads evidence supporting ideas contrary to one's personal opinions;

(d) Accepts no conclusions as final or ultimate.

3. Bases opinions and conclusions on adequate evidence

(a) Is slow to accept as facts anything not supported by convincing proof;

(b) Bases conclusions upon evidence obtained from a variety of dependable sources;

(c) Searches for the most satisfactory explanation of observed phenomena that the evidence permits;

(d) Sticks to the facts and refrains from exaggeration;

(e) Does not permit personal pride, bias, prejudice, or ambition to prevent the truth;

(f) Does not make snap judgements or jump to conclusions.

4. Evaluates techniques and procedures used and information obtained

(a) Uses a planned procedure in solving his problems;

(b) Seeks to use the various techniques which have proved valuable in obtaining evidence;

(c) Seeks to adopt the various techniques and procedures to the problem at hand;

(d) Makes the scientist personally consider the evidence and to decide whether it relates to the problem, or not;

(e) Judges whether the evidence is sound, sensible and complete enough to allow a conclusion to be drawn;

(f) Selects the most recent authoritative and accurate evidence related to the problem.

5. Is curious concerning the things observed

(a) Wants to know the "whys", "whats", and "hows" of observed phenomena;

(b) Is not satisfied with vague explanations of questions.

On the other hand, the Fifty-ninth Year Book of the National Society for the Study of Education (1960) condensed the list of habits constituting Scientific Attitudes into five. According to it, a person with Scientific Attitude-

(1) looks open-minded towards the work and opinion of others and towards information related to his problem;

(2) bases opinions and conclusions on adequate evidence;

(3) evaluates techniques and procedures used and information obtained;

(4) is curious concerning things he observes; and

(5) Looks for the natural causes of events.

Haney, R.E. (1964) proposed eight aspects of Scientific Attitude which are as follows:

1. Curiosity
2. Rationality
3. Willingness to suspend judgement
4. Open-mindedness
5. Critical-mindedness
6. Objectivity and intellectual honesty
7. Humility
8. Reverance for life

Diederich (1967) stated that Scientific Attitude constitutes twenty components. They are:

1. Scepticism
2. Faith in the possibility of solving problems
3. Desire for experimentation
4. Precision
5. A liking for new things
6. Willingness to change opinions
7. Humility
8. Loyalty to truth
9. An objective attitude
10. Aversion to superstition
11. Liking for scientific explanations
12. Desire for completeness of knoweldge
13. Suspended judgement
14. Ability to distinguish between hypotheses and solutions

15. Awareness of assumptions
16. Judgement of what is of fundamental and general significance
17. Respect for theoretical structures
18. Respect for quantification
19. Acceptance of probabilities, and
20. Acceptance of warranted generalisations.

Victor Y. Billeh and George A. Zakhariades (1975), on the basis of the definition made by Haney and Diederich, evolved a new definition constituting six components of Scientific Attitude. They have further described each behavioural aspect of the dimension of the Scientific Attitude. The six dimensions are as follows:

1. Rationality

(a) Commitment to the value of rationality.

(b) Tendency to test traditional beliefs.

(c) Seeking for natural causes of events and identification of cause-and-effect relationships.

(d) Acceptance of criticalness.

(e) Challenge of authority.

2 . Curiosity

(a) Desire for understanding new situations that are not explained by the existing body of knowledge.

(b) Seeking to find out the "whys" and "hows" of observed phenomena.

(c) Giving emphasis on the questioning approach for natural situations.

(d) Desire for completeness of knowledge.

3. Open-mindedness

(a) Willingness to revise opinions and conclusions.

(b) Desire for new things and ideas.

(c) Rejection of singular and rigid approach to people, things and ideas.

4. Aversion to Superstitions

(a) Rejection of Superstitious beliefs.

(b) Acceptance of Scientific facts and explanations.

5. Objectivity-Intellectual Honesty

(a) Demonstration of the greatest possible concern for observing and recording facts without any influence of personal pride, bias or ambition.

(b) In interpreting results, it does not allow any modification according to present social, economic, or political conditions.

6. Suspended Judgement

(a) Unwillingness to draw inferences before evidence is collected.

(b) Unwillingness to accept as facts things that are not supported by convincing proof.

(c) Avoidance of quick judgements and jumping to conclusions.

Through these six components, Billeh and Zakhariades constructed and standardised a thirty-six item Scientific Attitude Scale on Thurstone-Chave method. The purpose of the study was to (i) compare the Scientific Attitudes of secondary school pupils, university students and science teachers, and (ii) to correlate Scientific Attitudes of secondary school pupils with their achievement in science among 349 secondary school pupils, 121 University students and 31 science teachers. The results revealed that-

1. All the groups of subjects—secondary school pupils, University students and science teachers demonstrated positive scientific attitude.
2. The obtained differences among the mean scores on Scientific Attitude Scale of the five groups reveal that the amount of science knowledge or the general exposure to science courses affects positively the scientific attitudes.
3. It seems that the courses the students had from twelfth grade to sophomore did not influence the scientific attitudes. On the other hand, the science courses at the University Junior and Senior levels appeared to be effective in developing positive changes in the scientific attitudes of the students.
4. Since no significant difference was found between University students and science teachers one may conclude that experience in teaching secondary school science does not seem to be effective in developing further the scientific attitudes.
5. A low positive relationship exists between the scores on the Scientific Attitude Scale and the grades acheived by secondary school students in science.

Srivastava, N.N. (1980) constructed a thirty-six item Scientific Attitude Scale in Hindi constituting six dimensions, namely, rationality, curiosity, open-mindedness, aversion to superstitious beliefs, objectivity-intellectual honesty and suspended judgement and administered on a random sample of fifty science teachers, fifty non-science teachers, 100 science students and 100 non-science students of Madhya Pradesh. The main findings of his study were:

(i) The amount of scientifc knowledge or general exposure to science courses made impact on scientific attitude positively.

(ii) Scientific knowledge helped in the formation of scientific attitude.

(iii) Boys and Girls differed in respect of scientific attitude.

(iv) Male teachers and Female teachers did not differ in respect of scientific attitude.

Identification of the Dimensions and Sub-dimensions

The review of the literature, presented revealed that some of the components of Scientific Attitudes discussed by various authors are overlapping and some of them seem to be independent and totally discrete. As such, it was thought advisable to construct and standardise a comprehensive attitude scale for the purpose of the present study. And so, the components considered for the measurement of Scientific Attitude by various authors which are mentioned below were taken into consideration to identify the dimensions and their respective sub-dimensions for the construction of the present scale.

(a) Billeh and Zakhariade's definition of Scientific Attitude consisting of six dimensions with 19 sub-dimensions which seek to measure the behavioural aspect of dimensions.

(b) Haney's definition consisting of eight dimensions.

(c) Five aspects of Scientific Attitude which was spelt out by the Fifty-ninth Year Book of the National Society for the Study of Education.

(d) Noll's definition of Scientific Attitude consisting of six dimensions.

(e) The definition of the Scientific Attitude as defined in the Forty-sixth Year Book consisting of five dimensions with twenty sub-dimensions probing the behavioural aspect of the respective sub-dimensions; and

(f) Diederich's definition of Scientific Attitude constituting of twenty dimensions.

To identify the new dimensions and their sub-dimensions, the behavioural aspect of each of the dimension of the each of the said authors were typed on separate cards and were allotted inscriptions to the said authors as X_1, X_2, X_3, X_4, X_5, and X_6 respectively. Thus, X_1 cards in total were 19 which refer to the definition of Billeh and Zakhariades, in which X_1 1(a) to X_1 1(e) the five cards belonged to the dimensions of Rationality. Like-wise X_1 6(a) to X_1 6(c) the three cards belonged to the dimension of Suspended Judgement. In this manner X_2 cards from X_2 (a) to X_2 (h) of Haney's definition were prepared. Like-wise X_3, X_4, X_5 and X_6 cards were prepared pertaining to the authors mentioned in the preceding paragraphs under (c), (d), (e) and (f) respectively. Altogether, 78 cards in three sets were prepared. To finalise the dimensions and sub-dimensions, these three sets of cards were referred to the five experts in the field of education, two were from science background, two from social science background and one from psychology background. They were asked to pile the rest of the cards keeping the X_1 cards as the bases. While doing so, they were requested to—(i) pile the cards at one place which were identified to that of the behaviourial aspect of the each of the X_1 cards, and (ii) to keep the cards which were independent and discrete, separately.

After the evaluation of the each of the piles made by the experts, it was thought advisable to add a new dimension to the Scientific Attitude, namely, Theory Building with the following four behavioural aspects:

(i) Awareness of assumptions,

(ii) Respect for theoretical structures,

(iii) Acceptance of probabilities, and

(iv) Acceptance of warranted generalisations.

And further, in view of the Judges' opinions the following were added to the main dimensions:

1. Rationality

(a) humility

3. Open-mindedness

(a) Believes that truth never changes, but his ideas about what is true may change as he gains better understanding of that truth.

(b) Listens to, observes, or reads evidences supporting ideas contrary to his personal opinion.

5. Objectivity-Intellectual Honesty

(a) Searches for the most satisfactory explanation of observed phenomena that the evidence permits.

(b) Judges whether the evidence is sound, sensible and complete enough to allow a conclusion to be drawn.

(c) Selects the most recent authoritative and accurate evidence related to the problem.

(d) Accuracy-Precision.

(e) Respect for quantification.

Finalisation of the Dimensions and Sub-dimensions

Thus, the components of Scientific Attitude as evolved by the present investigator are as follows:

1. Rationality

(a) Commitment to the value of Rationality.

(b) Tendency to test traditional beliefs.

(c) Seeking for Natural causes of events and identification of cause-and-effect relationships.

(d) Acceptance of Criticalness.

(e) Challenge of Authority.

(f) Humility.

2. *Curiosity*

(a). Desire for understanding new situations.

(b) Seeking to findout the whys and hows of observed phenomena.

(c) Giving emphasis to questioning approach.

(d) Desire for completeness of knowledge.

3. *Open-mindedness* .

(a) Willingness to revise opinions and conclusions.

(b) Desire for new things and ideas.

(c) Rejection of singular and rigid approach to people, things and ideas.

(d) Believes that truth never changes, but his ideas of what is true may change as he gains better understanding of that truth.

(e) Listens to, observe or reads, evidence supporting ideas contrary to his personal opinions.

4. *Aversion to Superstitious Beliefs*

(a) Aversion to superstitious beliefs.

(b) Acceptance of scientific facts and explanations.

5. *Objectivity-Intellectual Honesty*

(a) Demonstration of the greatest possible concern for observing and recording of facts.

(b) In interpreting results, not allowing any modifications according to present social, economic and political situations.

(c) Searches for the most satisfactory explanation of observed phenomena that the evidence permits.

(d) Judges whether the evidence is sound, sensible and complete enough to allow a conclusion to be drawn.

(e) Selects the most recent authoritative and accurate evidence related to the problem.

(f) Accuracy-Precision.

(g) Respect for quantification.

6. Suspended Judgement

(a) Unwillingness to draw inferences before evidence is collected.

(b) Unwillingness to accept as facts, things that are not supported by convincing proof.

(c) Avoidance of quick judgements and jumping to conclusions.

7. Theory Building

(a) Awareness of Assumptions.

(b) Respect for theoretical structures.

(c) Acceptance of probabilities.

(d) Acceptance of warranted generalisations.

Construction of the Statements

Thus, the present definition of Scientific Attitude includes seven dimensions and thirty-one sub-dimensions with the behavioural aspects of scientific attitude. On the behavioural aspects of thirty-one dimensions, ninety statements of Scientific Attitude were constructed keeping in view the related literature. Out of which thirty-nine statements were positive and fifty-one statements, negative. Care has been taken to construct the parallel forms of statements in equal number on positive and negative sides. Thus, the statements so constructed positive and negative, were parallel to each other. These statements were typewritten on three sets and were referred to three judges to scrutinise and refine the statements whereever necessary in the light of the behavioural aspect of each dimension. In this process some statements were added, some were dropped and some were trimmed. Further, the judges were asked to select 31 positive statements and 31 negative statements. The 62 statements thus selected by the judges were jumbled and were printed in the form of a Book-Let for the purpose of standardisation of a Scientific Attitude Scale on "Thurstone-Chave method of equal appearing intervals". The statements which were printed on the Book-lets are as follows:

Unfavourable Natural Favourable

1. I accept only those things which stand the test of reasoning.

A	B	C	D	E	F	G	H	I	J	K

2. Properly conceived theoretical generalizations need not be changed as the new generalizations emerge.

A	B	C	D	E	F	G	H	I	J	K

3. I shall not accept any tradition; I belief unless I test it.

A	B	C	D	E	F	G	H	I	J	K

4. When correct theorization is made, the probabilities get reduced to one.

A	B	C	D	E	F	G	H	I	J	K

5. Any natural calamity is the result of certain natural causes and so to understand the nature of calamity, its antecedents need to be studied.

A	B	C	D	E	F	G	H	I	J	K

Unfavourable Natural Favourable

6. Raising theoretical structures is a fanciful romanticism.

A	B	C	D	E	F	G	H	I	J	K

7. Critical faculties are instrumental in expanding the horizons of knowledge.

A	B	C	D	E	F	G	H	I	J	K

8. Science is based on concrete facts, not on assumptions.

A	B	C	D	E	F	G	H	I	J	K

9. No authority is infalliable. So one should not hesitate to challenge it.

A	B	C	D	E	F	G	H	I	J	K

10. One should not hesitate to take fast judgements and drawing conclusions.

A	B	C	D	E	F	G	H	I	J	K

11. Humility does not signify confession of ignorance. Even when one is sure that he knows more than others, he need not give up humility.

A	B	C	D	E	F	G	H	I	J	K

12. One need not insist for supporting proof when great people themselves make statement of facts.

A	B	C	D	E	F	G	H	I	J	K

13. As the old order changes yielding place to new, one needs to be always ready to understand the new.

A	B	C	D	E	F	G	H	I	J	K

14. It is obvious that often we draw final inference first and then collect supporting evidence.

A	B	C	D	E	F	G	H	I	J	K

Unfavourable Natural Favourable

Statement	A	B	C	D	E	F	G	H	I	J	K
15. I find it difficult to resist my curiosity to find out the whys and hows of observed phenomena.											
16. Quantification which is being employed these days is an untrustworthy mechanistic device.											
17. If I were a teacher I would encourage my student to question me as closely as they can.											
18. Perfect precision is a dream. When observations are made through human medium which is far from perfect, it is unrealistic to aim at perfect precision.											
19. The scientists unending quest is a part of his desire to understand the reality in its totality.											
20. Old evidences related to a problem are more authentic than the most recent ones.											
21. As I grow I often find myself revising my opinions and conclusions.											
22. One need not rule out the possibility of a lag between the drawing of conclusion and completeness of evidence.											

Unfavourable **Natural** **Favourable**

23. I feel excited when I see a new thing or hear a new idea.

A	B	C	D	E	F	G	H	I	J	K

24. When the observed phenomenon is self evident, one need not insist for further explanation.

A	B	C	D	E	F	G	H	I	J	K

25. I do not feel uncomfortable in revising the approaches I formulate towards people, things and ideas.

A	B	C	D	E	F	G	H	I	J	K

26. In interpreting results one has to give sufficient latitude to present social, economic and political situation.

A	B	C	D	E	F	G	H	I	J	K

27. It is likely that much of the scientific information we have today will be demonstrated to be inaccurate or inadequate in the future.

A	B	C	D	E	F	G	H	I	J	K

28. It is the mediocre who insists on observation and recording of facts as the wise can formulate conclusions even without the crude data.

A	B	C	D	E	F	G	H	I	J	K

29. I do not hesitate to receive any evidence, writen or spoken which opposes my strongest personal opinion.

A	B	C	D	E	F	G	H	I	J	K

Statement	A	B	C	D	E	F	G	H	I	J	K
	Unfavourable					Natural					Favourable
30. When the findings or theories of Science conflict with religious belief, it is better to accept the religious belief.											
31. I hate all superstitions which mosquarade as traditions in our country.											
32. There are many things in the world which transcend human rationality and so must be accepted.											
33. I prefer to accept facts which are rationally explained.											
34. The traditional views have survived over the centuries because of their validity and so they must be accepted.											
35. A good scientist always demonstrates concern for observation and recording of facts.											
36. Gandhiji has asserted that the Bihar earth quake was a punishment inflicted by the nature on Biharis for the sin of untouchability.											
37. The vagaries of present social, economic and political situation should not modify the interpretation of results.											

Unfavourable **Natural** **Favourable**

Statement	A	B	C	D	E	F	G	H	I	J	K
38. Being always critical is not a virtue.											
39. Seeking further information about a novel situation is considered as a sound approach.											
40. Some, because of their expertise and mastery over a field of human activity are considered to be the authorities. It would be unwise to challenge them.											
41. Conclusions could be drawn only when the evidence is sensible and complete.											
42. It is hypocracy to demonstrate humility when one is sure that he knows more than others.											
43. When a gamut of authoritative evidence relating to a problem are furnished, I prefer to select the most recent one.											
44. All the possible patterns are already conceived by the great theoreticians I adore. Nothing new which has not already been theorised can emerge which merits renewed understanding.											

	Unfavourable					Natural					Favourable
	A	B	C	D	E	F	G	H	I	J	K
45. Accuracy and precision are the cornerstones of Science.											
46. We must be concerned with the phenomenon we confront with. It is sheer adventurism to seek to unearth the antecedents of each observed phenomenon.											
47. No meaningful conclusions can be drawn unless the evidence is quantified.											
48. When candid and lucid narrations can make the students properly understand the things to be taught, it is (naive) pedantic to resort to questioning method.											
49. I do not feel like drawing final inferences before convincing evidence is collected.											
50. Man has already the command over the totality of knowledge. Still greater efforts in the directions of knowing more are uncalled for.											
51. One need not accept the statements emanating from great men as facts unless they are supported by convincing proof.											

Unfavourable **Natural** **Favourable**

52. As I am not in the habit of entertaining wrong opinions and defective conclusions. I need not change them.

A	B	C	D	E	F	G	H	I	J	K

53. I feel hesitant to take snap judgements when the evidence does not seem to warrant such fast judgements.

A	B	C	D	E	F	G	H	I	J	K

54. I do not feel comfortable when I confront things and ideas which I did not come across earlier.

A	B	C	D	E	F	G	H	I	J	K

55. Scientific Theory is based on certain assumptions. The moment the assumptions are proved wrong, the entire citadel of science collapses.

A	B	C	D	E	F	G	H	I	J	K

56. When I am convinced that the approach I have adopted is perfect why should I adopt multiple alternative approaches to people, things and ideas.

A	B	C	D	E	F	G	H	I	J	K

57. The Theoretical structures which are raised so strenuously need to be respected.

A	B	C	D	E	F	G	H	I	J	K

58. As the Truth is unchanging, the ideas about the Truth should remain unchanged.

A	B	C	D	E	F	G	H	I	J	K

Unfavourable Natural Favourable

59. There are always alternative competing probabilities which deserve to be accepted.

A	B	C	D	E	F	G	H	I	J	K

60. Obviously I do not feel very comfortable when I listen to observe or read evidence contradicting my strong personal opinions.

A	B	C	D	E	F	G	H	I	J	K

61. When evidence warrants a new generalization, one should not hesitate to shed the old and accept the new generalization.

A	B	C	D	E	F	G	H	I	J	K

62. Even the so called superstition has deep roots in reality which we often fail to take into cognisance.

A	B	C	D	E	F	G	H	I	J	K

Fixing the Intensity of the Statements

These booklets containing draft statements were presented to a panel of 33 judges who were experts in the field of education. They belonged to various institutions and were drawn from different faculties. The panel of judges comprised 13 Professors, 10 Readers, 9 Lecturers and One Engineer. They were requested to go through a statement first to decide whether it was a positive or a negative statement and then to fix the intensity of the statement towards positiveness or negativeness depending upon the nature of the statement in question on an eleven point scale arranged in eleven boxes against each statement with the superscription of boxes from "A" to "K" for each statement. The intensity of the positiveness of the statement towards measuring Scientific Attitude increased from "F" to "K", and the intensity of the negativeness of the statement for measuring Scientific Attitude increased from "F" to "A". While "A" stood for most unfavourable position of the statement, "F" stood for the neutral position and "K" for the most favourable position of the statement. In this way, the judges were asked to fix the intensity/weightage of each statement on the scale in a continuum form "A" to "K" with a (X) mark in the box.

Selection of the Statements

The data so obtained from 33 judges were subjected to statistical treatment to get the scale values (Median value) and 'Q' values (semi-inter quartile range) for each statement by plotting a Cumulative Frequency Graph. In view of paucity of space only a few Cumulative Frequency Graphs showing the Scale values and 'Q' values of certain items are presented in the following pages. However, the entire data pertaining to each statement is furnished in Appendics 3 in the order of items detailed in the booklet. The Scale value, S, of each statement was obtained by the median of the distribution of judgements for each statement. The 'Q' value was found by calculating the interquartile range of the distribution of the judgements for each statement.

Cumulative Frequency Graph for Item No. 1

I accept only those things which stand the test of reasoning

Pile No.	*Frequency*	*Percentage*	*Cumulative Percentage*
1.	0	0	0
2.	0	0	0
3.	0	0	0
4.	0	0	0
5.	0	0	0
6.	2	6.06	6.06
7.	1	3.03	9.09
8.	5	15.15	24.24
9.	2	6.06	30.30
10.	7	21.21	51.51
11.	16	48.49	100.00

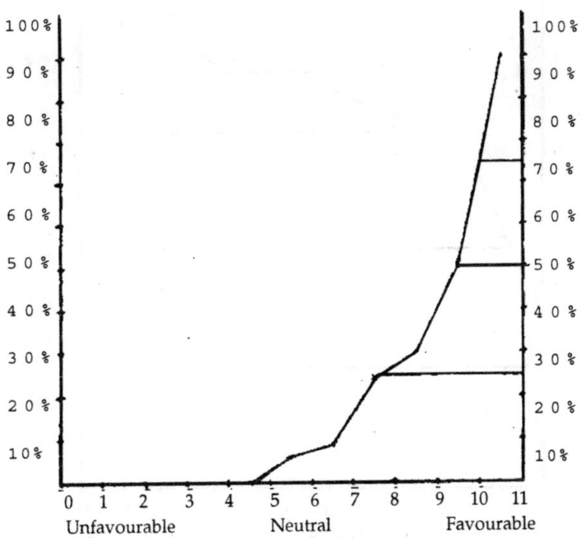

Median = 9.5 Q=1/2 (10–7.6) = 1.2 Item Selected

Cumulative Frequency Graph for Item No. 12

One need not insist for supporting proof when great people themselves make statement of facts

Pile No.	*Frequency*	*Percentage*	*Cumulative Percentage*
1.	15	45.46	45.46
2.	7	21.21	66.67
3.	2	6.06	72.73
4.	2	6.06	78.79
5.	1	3.03	81.82
6.	2	6.06	87.88
7.	0	0	87.88
8.	0	0	87.88
9.	2	6.06	93.94
10.	1	3.03	96.97
11.	1	3.03	100.00

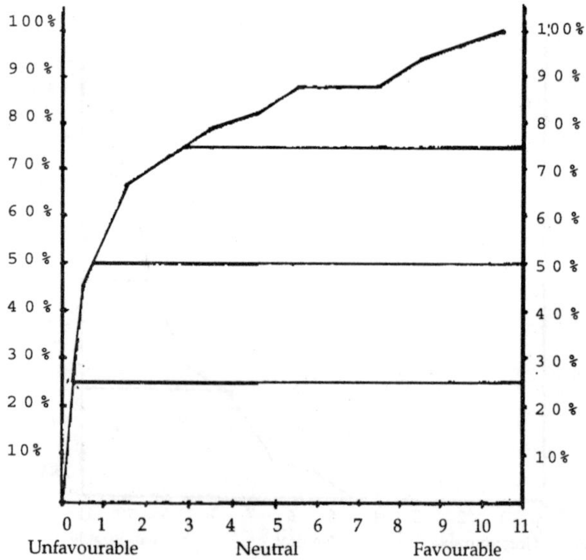

Median = 0.70 Q=1/2 (2.9–0.3) = 1.30 Item Selected

Cumulative Frequency Graph for Item No. 18

Perfect precision is a dream, when observations are made through human medium which is far from perfect, it is unrealistic to aim at perfect precision

Pile No.	*Frequency*	*Percentage*	*Cumulative Percentage*
1.	1	3.23	3.23
2.	0	0	3.23
3.	0	0	3.23
4.	1	3.23	6.46
5.	1	3.23	9.69
6.	4	12.90	22.59
7.	2	6.45	29.04
8.	0	0	29.04
9.	9	29.03	58.07
10.	9	29.03	87.10
11.	4	12.90	100.00

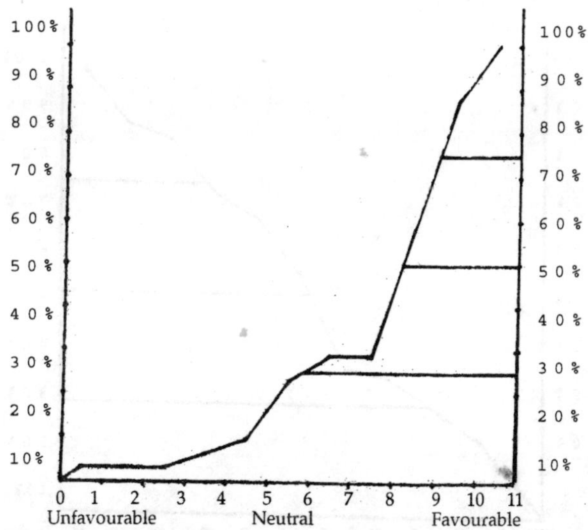

Median = 8.20 Q=1/2 (9.1–5.8) = 1.65 Item Selected

But Eliminated

Cumulative Frequency Graph for Item No. 38

Being always critical is not a virtue

Pile No.	*Frequency*	*Percentage*	*Cumulative Percentage*
1.	2	6.06	6.06
2.	4	12.12	18.18
3.	2	6.06	24.24
4.	2	6.06	30.30
5.	1	3.03	33.33
6.	7	21.21	54.54
7.	4	12.12	66.66
8.	2	6.06	72.72
9.	4	12.12	84.84
10.	1	3.03	87.87
11.	4	12.12	100.00

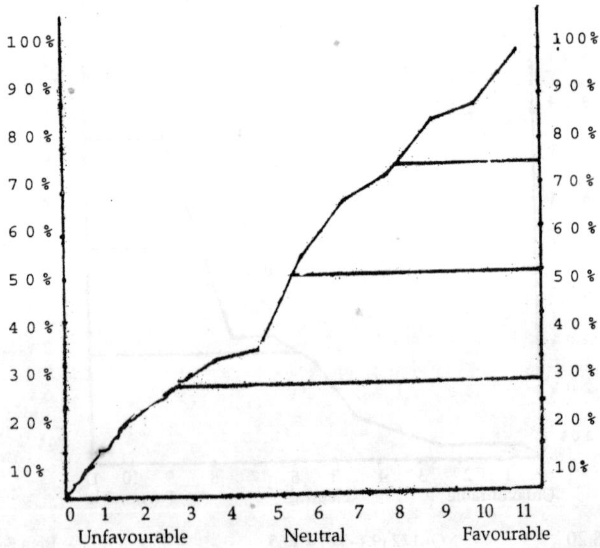

Median = 5.30 Q=1/2 (7.7–2.7) = 2.50 Item Selected

Cumulative Frequency Graph for Item No. 41

Conclusions could be drawn only when the evidence is Sensible and Complete

Pile No.	*Frequency*	*Percentage*	*Cumulative Percentage*
1.	0	0	0
2.	0	0	0
3.	0	0	0
4.	1	3.03	3.03
5.	1	3.03	6.06
6.	1	3.03	9.09
7.	4	12.12	21.21
8.	4	12.12	33.33
9.	8	24.24	57.57
10.	4	12.12	69.69
11.	10	30.30	100.00

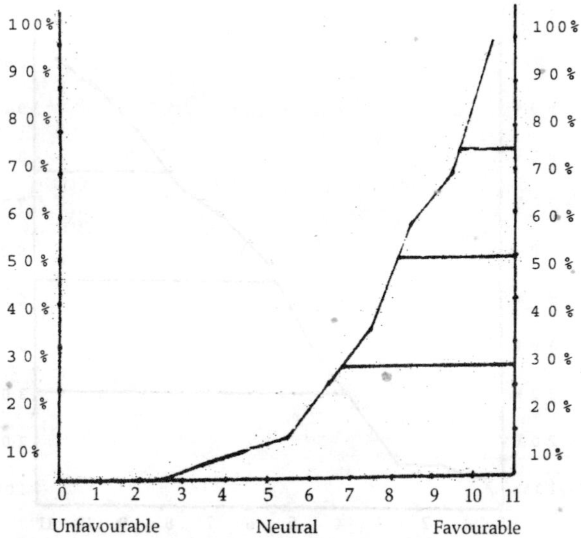

Median = 8.20 Q=1/2 (9.7–6.9) = 1.40 Item Selected

But Eliminated

Cumulative Frequency Graph for Item No. 43

When a gramut of authoritative evidence relating to a problem are furnished, I prefer to select the most recent one

Pile No.	*Frequency*	*Percentage*	*Cumulative Percentage*
1.	2	6.45	6.45
2.	1	3.23	9.68
3.	0	0	9.68
4.	4	12.90	22.58
5.	4	12.90	35.48
6.	6	19.36	54.84
7.	3	9.68	64.52
8.	2	6.45	70.97
9.	4	12.90	83.87
10.	3	9.68	93.55
11.	2	6.45	100.00

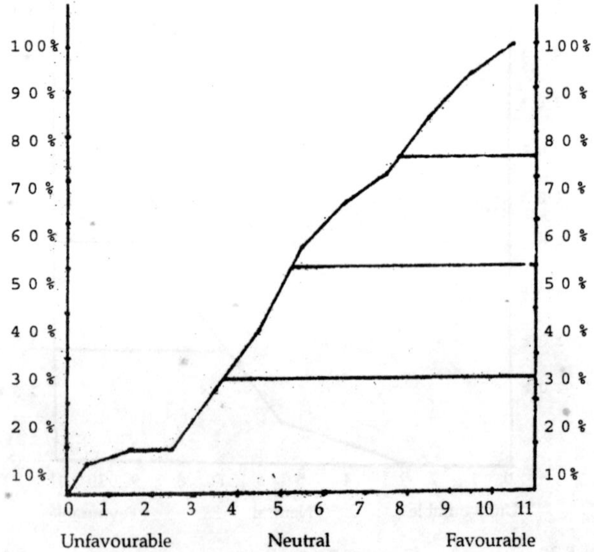

Median = 5.20 Q=1/2 (7.8 –3.7) = 2.50 Item Selected

Cumulative Frequency Graph for Item No. 51

One need not accept the statements emanating from great men as facts unless they are supported by convincing proof

Pile No.	*Frequency*	*Percentage*	*Cumulative Percentage*
1.	1	3.03	3.03
2.	1	3.03	6.06
3.	0	0	6.06
4.	1	3.03	9.09
5.	1	3.03	12.12
6.	0	0	12.12
7.	1	3.03	15.15
8.	5	15.15	30.30
9.	2	6.06	36.36
10.	10	30.30	66.66
11.	11	33.33	100.00

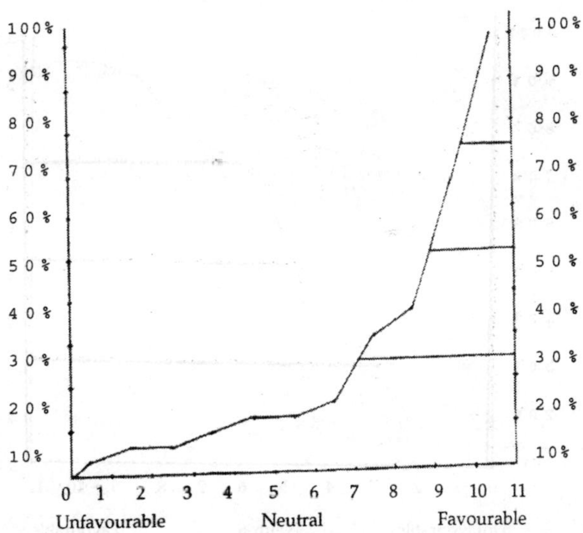

Median =9.00 Q=1/2 (9.7–7.1) = 1.30 Item Selected

Cumulative Frequency Graph for Item No. 62

Even the so called superstitions has deep roots in reality which we often fail to take into congnisance

Pile No.	*Frequency*	*Percentage*	*Cumulative Percentage*
1.	4	12.90	12.90
2.	1	3.23	16.13
3.	3	9.68	25.81
4.	1	3.23	29.04
5.	3	9.68	38.72
6.	10	32.26	70.98
7.	2	6.45	77.53
8.	2	6.45	83.88
9.	4	12.90	96.78
10.	1	3.23	100.00
11.	0	0	100.00

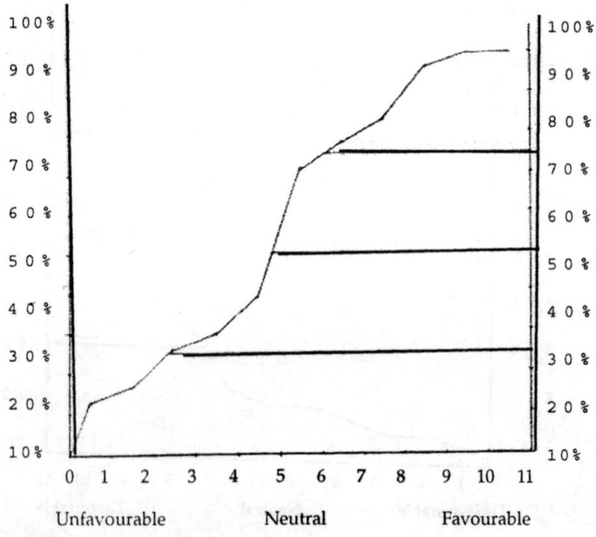

Median = 4.80 Q=1/2 (6.1-2.5) = 1.80 Item Selected

On the basis of the S and Q values, out of 62 statements, 48 statements - 24 positive and 24 corresponding negative statements-were selected for the present scale. The scale value of the selected statements covered the range of the continuum and had low Q values. Statements with Q values above 3.00 were eliminated. And the statements which were constructed parallel to them were also eliminated to keep the uniformity in selecting the items for present scale. The statements which were scaled at one end of the continuum-irrespective of its positiveness and negativeness and vice versa-were eliminated inorder to avoid ambiguity, though they had low Q values. Thus, after the standardisation process 48 statements which emerged with S and Q values are as follows:

Scientific Attitude Scale

	Scale Value	*Q-Value*
1. I accepted only those things which stand the test of reasoning	9.50	1.20
2. When I am convinced that the approach I have adopted is perfect why should I adopt multiple alternative approaches to people, things and ideas.	2.70	1.70
3. As the Truth is unchanging, the ideas about the Truth should remain unchaged.	2.30	1.80
4. I shall not accept any traditional belief unless I test it.	8.00	2.00
5. Obviously, I do not feel very comfortable when I listen to; observe or read evidence contradicting my strong personal opinions.	3.50	2.60
6. Any natural calamity is the result of certain natural causes and so to		

	Scale Value	Q-Value
understand the nature of calamity, its antecedents need to be studied.	9.20	0.80
7. Critical faculties are instrumental in expanding the horizons of knowledge.	9.50	0.65
8. Even the so called superstitions have deep roots in reality which we often fail to take into cognisance.	4.80	1.80
9. No authority is infallible. So one should not hesitate to challenge it.	9.40	0.65
10. When the findings or theories of Science conflict with religious belief, it is better to accept the religious belief.	1.10	1.20
11. It is the mediocre who insists on observation and recording of facts as the wise can formulate conclusions even without the crude data.	1.90	1.75
12. Humility does not signify confession of ignorance. Even when one is sure that he knows more than others, he need not give up humility.	9.00	1.65
13. When the observed phenomenon is self evident, one need not insist for further explanation.	4.20	2.50
14. As the old order changes yielding place to new, one needs to be always ready to understand the new.	8.80	0.95
15. I find it difficult to resist my curiosity to find out the whys and hows of observed phenomena.	8.90	0.75
16. Old evidences related to a problem are more authentic than the most recent ones.	2.40	1.15

	Scale Value	Q-Value
17. If I were a teacher I would encourage my students to question me as closely as they can.	10.00	0.30
18. Quantification which is being employed these days is an untrustworthy mechanistic device.	2.40	1.40
19. It is obvious that often we draw final inference first and then collect supporting evidence.	4.70	2.25
20. The Scientist's unending quest is a part of his desire to understand the reality in its totality.	9.00	0.80
21. One need not insist for supporting proof when great people themselves make statement of facts.	0.70	1.30
22. As I grow I often find myself revising my opinions and conclusions.	8.30	1.00
23. I feel excited when I see new things or hear a new idea.	8.30	1.25
24. When correct theorization is made, the probabilities get reduced to one.	5.40	2.05
25. I do not feel uncomfortable in revising the approaches I formulate towards people, things and ideas.	8.20	1.10
26. There are many things in the world which transcend human rationality and so must be accepted.	4.50	2.20
27. The traditional views have survived over the centuries because of their validity and so they must be accepted.	3.20	2.25
28. It is likely that much of the information we have today will be		

	Scale Value	*Q-Value*
demonstrated to be in-accurate or inadequate in the future.	6.30	2.65
29. Gandhiji has asserted that the Bihar earth quake was a punishment inflicted by the nature on Biharies for the sin of untouchability.	0.50	1.35
30. I do not hesitate to receive any evidence, written or spoken, which opposes my strongest personal opinion.	8.80	1.00
31. I hate all superstitions which mosquerade as traditions in our country.	7.50	2.70
32. Being always critical is not a virtue.	5.30	2.50
33. I prefer to accept facts which are rationally explained.	9.10	1.25
34. Some, because of their expertise and mastery over a field of human activity, are considered to be the authorities. It would be unwise to challenge them.	1.70	1.55
35. It is hypocracy to demonstrate humility when one is sure that he knows more than others.	4.70	1.95
36. A good scientist always demonstrates concern for observation and recording of facts.	9.40	0.80
37. All the possible patterns are already conceived by the great theoreticians I adore. Nothing new which has not already been theorised can emerge which merits renewed understanding.	0.60	1.35
38. Seeking further information about a novel situation is considered as a sound approach.	8.20	2.25

	Scale Value	Q-Value
39. When a gamut of authoritative evidence relating to problem is furnished, I prefer to select the most recent one.	5.20	2.05
40. We must be concerned with the phenomenon we confront with. It is sheer adventurism to seek to unearth the antecedents of each observed phenomenon.	4.20	2.10
41. No meaningful conclusions can be drawn unless the evidence is quantified.	7.10	2.45
42. When candid and lucid narrations can make the students properly understand the things to be taught, it is (naive) pedantic to resort to questioning method.	3.30	1.85
43. Man has already the command over the totality of knowledge. Still greater efforts in the directions of knowing more are uncalled for.	0.40	0.65
44. I do not feel like drawing final inferences before convincing evidence is collected.	8.70	1.20
45. As I am not in the habit of entertaining wrong opinions and defective conclusions, I need not change them.	2.00	1.40
46. One need not accept the statements emanating from great men as facts unless they are supported by convincing proof.	9.00	1.30

	Scale Value	*Q-Value*
47. There are always alternative competing probabilities which deserve to be accepted.	8.10	1.25
48. I do not feel comfortable when I confront things and ideas which I did not come across earlier.	4.70	2.75

Dimensions and Sub-dimensions of the Present Scale

The 48 statements (24 positive and 24 negative statements which are arranged in jumbled manner) thus measure the following dimensions and sub-dimensions of the Scientific Attitude for the present study.

1. Rationality

(a) Commitment to the value of Rationality, (measured by items - No. 1 & No. 26)

(b) Tendency to test traditional beliefs, (measured by items - No. 4 & No. 27)

(c) Seeking for Natural causes of events and identification of cause-and-effect relationships, (measured by items - No. 6 & No. 29)

(d) Acceptance of criticalness. (measured by items - No. 7 & No. 32)

(e) Challenge of Authority. (measured by items - No. 9 & No.34)

(f) Humility. (measured by items - No. 12 & No.35)

2. Curiosity

(a) Desire for understanding new situations. (measured by items - No. 14 & No. 37)

(b) Seeking to findout the whys and hows of observed phenomena.
(measured by items - No. 15 & No. 40)

(c) Giving emphasis on questioning approach.
(measured by items - No. 17 & No. 42)

(d) Desire for completeness of knowledge.
(measured by items - No. 20 & No. 43)

3. *Open-Mindedness*

(a) Willingness to revise opinions and conclusions.
(measured by items - No. 22 & No. 45)

(b) Desire for new things and ideas.
(measured by items - No. 23 & No. 48)

(c) Rejection of singular and rigid approach to people, things and ideas.
(measured by items - No. 25 & No. 2)

(d) Believes that truth never changes, but his ideas of what is true may change as he gains better understanding of that truth.
(measure by items - No. 28 & No. 3)

(e) Listens to, observes or reads evidence supporting ideas contrary to his personal opinions.
(measured by items - No. 30 & No. 5)

4. *Aversion to Superstitious Beliefs*

(a) Aversion to superstitious beliefs.
(measured by items - No. 31 & No. 8)

(b) Acceptance of scientific facts and explanations.
(measured by items - No. 33 & No. 10)

5. *Objectivity-Intellectual Honesty*

(a) Demonstration of the greatest possible concern for observing and recording facts.
(measured by items - No. 36 & No.11)

(b) Searches for the most satisfactory explanation of observed phenomena that the evidence permits. (measured by items - No. 38 & No. 13)

(c) Selects the most recent authoritative and accurate evidence related to the problem. (measured by items - No. 39 & No. 16)

(d) Respect for quantification. (measured by items - No.41 & No. 18)

6. Suspended Judgement

(a) Unwillingness to draw inferences before evidence is collected. (measured by items - No.44 & No. 19)

(b) Unwillingness to accept as facts, things that are not supported by convincing proof. (measured by items No.46 & No. 21)

7. Theory Building

(a) Acceptance of probabilities. (measured by items - No.47 & No. 24)

Validity and Reliability of the Scale

The present scale, in the process of construction and standardisation has been given rigorous treatment. At various levels it was subjected to scrutiny by judges and their opinions were followed. The items which were selected had low Q values. The intercorrelations among the dimensions which were obtained through the scores of the present sample are presented in the Table 2.1.

The Table 2.1 reveals that by and large the correlations between the dimensions are positive and significant at 0.01 levels. Thus the dimensions of the present scale have intercorrelations. And so the present scale has its face validity, content validity and construct validity.

Reliability

The reliability of the scale measured by odd-even and split-half methods on the present sample were 0.70 and 0.63 respectively. The coefficient of correlations were very high and were significant at 0.01 level.

Thus the present scale is valid and reliable.

Table 2.1

Correlation of Each Dimension of Scientific Attitude with that of other Dimensions

	D1	*D2*	*D3*	*D4*	*D5*	*D6*	*D7*	*D8*
D1	1.0000	0.3803**	0.2663**	0.4113**	0.3303**	0.3687**	0.1200*	0.7318**
D2			0.4201**	0.2677**	0.4278**	0.3679**	0.1684**	0.7551**
D3				0.1240*	0.2388**	0.1970**	0.2451**	0.6243**
D4					0.2567**	0.3161**	0.0216	0.5271**
D5						0.3957**	0.0663	0.6511**
D6							0.0466	0.6020**
D7								0.2631**
D8								1.0000

** Significant at 0.01 level

* Significant at 0.05 level

D1=Rationality; D2= Curiosity; D3=Open-mindedness; D4=Aversion to Superstitious Beliefs;
D5 = Objectivity-Intellectual Honesty; D-6=Suspended Judgement; D7=Theory Building; D8=Total Dimensions

Thus the second chapter has been devoted to the presentation of the process of standardisation of the Scientific Attitude Scale. In the third chapter the results obtained through the analysis of the data collected from the Academics and Research Scholars from certain universities underlining the relationships between the various demographic factors of the sample and their association with the Scientific Attitude are presented.

3
Measurement of Scientific Attitude among Certain Intellectual Groups

The sole objective of the present study was to construct and standardise a Scientific Attitude Scale to measure the Scientific Attitude among the Academics and Research Scholars in Universities in relation to their demographic factors. The scale thus constructed and standardised for this purpose has been explained in the second chapter. Besides, the research envisaged to measure the Scientific Attitude of Academics in relation to certain demographic factors of the subjects, the demographic factors sought to be measured are put in part A of the scale (See Appendix). In the present chapter it is proposed to discuss the Part A of the scale, sample, data collection procedure adopted, statistical techniques used, Results and discussion of results under separate captions.

Tools

Keeping in view the objectives of the present study, the researcher has constructed and standardised a tool on Scientific Attitude Scale. The process of standardisation of the tool has been detailed in the second chapter.

The present tool consists of two parts. The first part covers the personal data bank which helps to record and measure the following variables:

(1) Sex

(2) Occupation

(3) Age

(4) Income

(5) Main content background

(6) Experience

(7) Type of qualification

(8) Political ideological background

(9) Level of Political activism

(10) Political commitment.

The subjects were asked to put a tick mark (√) in the appropriate boxes to record the said information. The second part of the tool consists of 48 statements that help to measure the Scientific Attitude of a person, which is called the "Scientific Attitude Scale". The present scale tries to measure the Scientifc Attitude of a person in totality and the dimension-wise Scientific Attitude as well. The subjects were to put a tick mark (√) in the appropriate boxes i.e., "Agree" and "Disagree" which were arranged against each statement of the Scientific Attitude. The data thus obtained was, given the ascribed scale values and then it was analysed statistically variable-wise and conclusions arrived at.

Sample

The present investigation drew its sample from the following four institutions:

(1) Osmania University, Hyderabad.

(2) Regional Engineering College, Warangal.

(3) Kakatiya University, Warangal.

(4) Kakatiya Medical College, Warangal.

The Academics and Research Scholars of the said four institutions formed the sample for the present study.

The cross section of the sample so obtained for the purpose of investigation is presented in the following paragraphs in different tables.

Table 3.1

Sample-Sexwise and Occupation Level-wise

Sex/Occupation	*Males*	*Females*	*Total*
Academics	317	52	369
Research Scholars	124	16	140
	441	68	509

Table 3.1. reveals that a total of 509 subjects formed the sample for the present investigation of which 441 were males and 68 females. Out of 509 subjects, 369 were academics and 140 research scholars.

One of the main objectives of the present investigation was to find out relationship between age and Scientific Attitude. Table 3.2. gives the agewise breakup of the sample. For the sake of convenience the entire sample has been divided into four age cohorts viz., 21-30 years, 31-40 years, 41-50 years and 51 years and above.

Table 3.2. reveals that out of 509 subjects, 503 responded to age column while 6 subjects left the age column blank. Out of 503 subjects, 140 subjects belonged to 21-30 years cohort, 140 subjects to 31-40 years age cohort, 147 subjects to 41-50 years age cohort and 76 subjects to 51 years and above age cohort.

Table 3.2

Sample-Agewise Breakup of the Sample

Age Cohort	*Number of subjects*
21-30 years	140
31-40 years	140
41-50 years	147
51 years and above	76
Total	503

Table 3.3

Sample-Income-wise Breakup of the Subjects

Income group	*Number of subjects*
Less than Rs. 500/- p.m.	22
Rs. 500 - Rs.1000/- p.m.	34
Rs.1000 - 2500/- p.m.	72
Rs. 2500 - 5000/- p.m.	189
Rs. 5000 - 8000/- p.m.	167
Rs. 8000 and above	8
Total	492

Table 3.3. reveals that out of total sample of 509 subjects, 492 subjects were getting income and the remaining 17 subjects were not getting any sort of income. Out of 492 who were getting income, 22 subjects were getting less than Rs.500/- per month; 34 subjects were getting Rs.500-1000/- per month; 72 subjects were getting income in the range of Rs.1000-2500/- per month, 189 subjects were getting income in the range of Rs. 2500-5000/- per month; 167 subjects were getting income in the range of Rs. 5000-8000/- per month and 8 subjects

were getting income in the range of Rs. 8000/- and above per month.

The present study intended to find out the relationship between the main content (subject) background of the subjects and their Scientific Attitude scores. The subject-background-wise sample is presented in the Table 3.4.

Table 3.4

Sample-Main Content Background-wise Breakup of the Sample

S.No.	*Main content background*	*Number of subjects*
1.	Arts and Humanities	71
2.	Social Sciences	95
3.	Physical Sciences	103
4.	Biological Sciences	66
5.	Health Sciences	86
6.	Engineering	58
7.	Mathematics	30
	Total	509

The Table 3.4 reveals that out of 509 total sample, 71 candidates belonged to the background of Arts and Humanities subjects, 95 candidates to Social Science subjects, 103 candidates to Physical Science subjects, 66 candidates to Biological Science subjects, 86 candidates to Health Science subjects; 58 candidates to Engineering and 30 candidates to Mathematics subject.

The present study equally intended to find out if the experience of the individual has an influence on the Scientific Attitude. The sample drawn in this regard is presented in the Table 3.5.

Table 3.5

Sample-Experience-wise Breakup of the Sample

Professional Experience	*Number of subjects*
Less than 5 years	106
06 - 10 years	92
11 - 15 years	73
16 - 20 years	53
21 - 25 years	55
26 - 30 years	33
31 - 35 years	9
Total	421

Table 3.5. reveals that out of total sample of 509 subjects, 421 were experienced. Out of 421 subjects with experience, 106 subjects were having 1-5 years of experience, 92 subjects were having 6-10 years of experience, 73 subjects were having 11-15 years of experience, 53 subjects were having 16-20 years of experience, 55 subjects were having 21-25 years of experience, 33 subjects were having 26-30 years of experience and 9 subjects were having 31-35 years of experience.

One of the main objectives of the study was to find out the relationship between the Scientific Attitudes of the subjects of Academic and Professional degree holders.

The relevant information covering the sample is presented in the Table 3.6.

Table 3.6 gives the details of the subjects who had acquired Academic and Professional degrees. Out of 509 subjects of the total sample 311 had acquired Academic degrees and 198 subjects had acquired Professional degrees.

Table 3.6

Sample-Academic and Professional Qualification-wise Breakup of the Sample

S.No.	Type of degree holders	Strength (N)
1.	Academic degree holders	311
2.	Professional degree holders	198
	Total	509

The present investigation was equally interested in finding out the relationships between the political ideology of the subjects and their Scientific Attitude.

The Political ideological background-wise sample is presented in the Table 3.7.

Table 3.7

Sample-Political Ideological-wise Breakup of the Sample

Political Ideology	No. of subjects who had background
R.S.S.	22
B. J .P	33
Congress	92
Janatadal	28
B.S.P.	5
C.P.I.	4
C.P.M.	14
C.P.I. (ML)	23
Any other	74
Total	295

Table 3.7 reveals that out of the total sample of 509 subjects, only 295 candidates informed that they had leanings towards

various Political ideologies. Among these 295, 22 subjects had leanings towards R.S.S., 33 towards B.J.P., 92 subjects towards Congress, 28 subjects towards Janatadal, 5 subjects towards B.S.P., 4 subjects towards C.P.I., 14 subjects towards C.P.M., 23 subjects towards C.P.I.(ML) and 74 subjects towards other Political ideologies.

The present investigation sought to find out the relationship between the level of Political activism and the Scientific Attitude. The related data obtained is presented in the Table 3.8.

Table 3.8

Sample-Political Activism Level-Breakup of the Sample

Level of Political activism	*Number of subjects*
Cadre/Worker Level	28
Organiser Level	21
Theoretician Level	104
Total	153

Table 3.8 shows that 153 out of 295 subjects who had Political Ideological leanings revealed their level of Political Activism. Out of 153 subjects who revealed level of activism, 28 were working in the Cadre/Worker level, 21 were working at the Organiser level and 104 were working at the Theoretician level.

The present investigation intended to find out the relationship between the Scientific Attitudes of the subjects who had background of some Political ideology and who had no background of any Political Ideology.

The relevant information covering the sample is presented in the Table 3.9.

Table 3.9

Sample-Politically Committed and Neutral

S.No.	*Type of Subjects*	*Strength (N)*
1.	Subjects who had background of some Political Ideology	295
2.	Subjects who had no background of any Political Ideology.	214
	Total	509

Table 3.9 reveals that out of the total sample of 509 subjects, 295 were having the background of some Political Ideology and 214 subjects were not having the background of any Political Ideology.

Data Collection Procedure

The present investigator visited the various departments in the following four institutions to collect the data in person:

(a) Osmania University, Hyderabad.

(b) Regional Engineering College, Warangal.

(c) Kakatiya University, Warangal.

(d) Kakatiya Medical College, Warangal.

The investigator spent 15-20 days at each institution in getting maximum number of Scientific Attitude Scale proformas filled in by the respondents of various faculties. On the First day, he would issue the questionnaires and explain the purpose of the study and collect the filled in proformas back after 2-3 days. Like this the data has been collected. He could collect data from 509 subjects even though he had issued scales to 700 subjects as the response was not total.

Statistical Techniques used

As it has already been stated the present investigation

sought to find out relationships between the following pairs of variables.

1. Sex and Scientific Attitude
2. Occupation and Scientific Attitude
3. Age and Scientific Attitude
4. Income and Scientific Attitude
5. Content Background and Scientific Attitude
6. Experience and Scientific Attitude
7. Type of Qualification and Scientific Attitude
8. Political Ideology and Scientific Attitude
9. Political Activism and Scientific Attitude
10. Political Commitment and Scientific Attitude.

The data so obtained from 509 subjects for the said purpose was statistically treated to obtain Scientific Attitude score of each individual.

The negative statement scale values were reflected upon the positive continuum inorder to have a uniform system of scale values for the positive and negative statements for proper statistical analysis. Thus after the process of reflection the scale values obtained for the entire scale is as follows:

	Scale Value
1. I accept only those things which stand the test of reasoning.	9.50
2. When I am convinced that the approach I have adopted is perfect why should I adopt multiple alternative approaches to people, things and ideas.	8.30
3. As the Truth is unchanging, the ideas about the Truth should remain unchanged.	8.70

	Scale Value
4. I shall not accept any traditional belief unless I test it.	8.00
5. Obviously I do not feel very comfortable when I listen to, observe, or read evidence contradicting my strong personal opinions.	7.50
6. Any natural calamity is the result of certain natural causes and so to understand the nature of calamity, its antecedents need to be studied.	9.20
7. Critical faculties are instrumental in expanding the horizons of knowledge.	9.50
8. Even the so called superstitions have deep roots in reality which we often fail to take into cognisance.	6.20
9. No authority is infallible. So one should not hesitate to challenge it.	9.40
10. When the findings or theories of Science conflict with religious belief, it is better to accept the religious belief.	9.90
11. It is the mediocre who insists on observation and recording of facts as the wise can formulate conclusions even without the crude data.	9.10
12. Humility does not signify confession of ignorance. Even when one is sure that he knows more than others, he need not give up humility.	9.00
13. When the observed phenomenon is self evident, one need not insist for further explanation.	6.80

	Scale Value
14. As the old order changes yielding place to new, one needs to be always ready to understand the new.	8.80
15. I find it difficult to resist my curiosity to find out the whys and hows of observed phenomena.	8.90
16. Old evidences related to a problem are more authentic than the most recent ones.	8.60
17. If I were a teacher I would encourage my students to question me as closely as they can.	10.00
18. Quantification which is being employed these days is an untrustworthy mechanistic device.	8.60
19. It is obvious that often we draw final inference first and then collect supporting evidence.	6.30
20. The scientists' unending quest is a part of his desire to understand the reality in its totality.	9.00
21. One need not insist for supporting proof when great people themselves make statement of facts.	10.30
22. As I grow I often find myself revising my opinions and conclusions.	8.30
23. I feel excited when I see a new thing or hear a new idea	8.30
24. When correct theorization is made, the probabilities get reduced to one.	5.60

	Scale Value
25. I do not feel uncomfortable in revising the approaches I formulate towards people, things and ideas.	8.20
26. There are many things in the world which transcend human rationality and so must be accepted.	6.50
27. The traditional views have survived over the centuries because of their validity and so they must be accepted.	7.80
28. It is likely that much of the scientific information we have today will be demonstrated to be inaccurate or inadequate in the future.	6.30
29. Gandhiji has asserted that the Bihar earth quake was a punishment inflicted by the nature on Biharies for the sin of untouchability.	10.50
30. I do not hesitate to receive any evidence, written or spoken, which opposes my strongest personal opinion.	8.80
31. I hate all superstitions which mosquerade as traditions in our country.	7.50
32. Being always critical is not a virtue.	5.70
33. I prefer to accept facts which are rationally explained.	9.10
34. Some, because of their expertise and mastery over a field of human activity are considered to be the authorities. It would be unwise to challenge them.	9.30
35. It is hypocracy to demonstrate humility when one is sure that he knows more than others.	6.30

	Scale Value
36. A good scientist always demonstrates concern for observation and recording of facts.	9.40
37. All the possible patterns are already conceived by the great theoreticians I adore. Nothing new which has not already been theorised can emerge which merits renewed understanding.	10.40
38. Seeking further information about a novel situation is considered as a sound approach.	8.20
39. When a gamut of authoritative evidence relating to a problem is furnished, I prefer to select the most recent one.	5.20
40. We must be concerned with the phenomenon we confront with. It is sheer adventurism to seek to unearth the antecedents of each observed phenomenon.	6.80
41. No meaningful conclusions can be drawn unless the evidence is quantified.	7.10
42. When candid and lucid narrations can make the students properly understand the things to be taught, it is (naive) pendantic to resort to questioning method.	7.70
43. Man has already the command over the totality of knowledge. Still greater efforts in the directions of knowing more are uncalled for.	10.60
44. I do not feel like drawing final inferences before convincing evidence is collected.	8.70

	Scale Value
45. As I am not in the habit of entertaining wrong opinions and defective conclusions, I need not change them.	9.00
46. One need not accept the statements emanating from great men as facts unless they are supported by convincing proof.	9.00
47. There are always alternative competing probabilities which deserve to be accepted.	8.10
48. I do not feel comfortable when I confront things and ideas which I did not come across earlier.	6.30

The maximum possible score on each dimension of the Scientific Attitude Scale is as follows:

Dimension	*Maximum Possible Score*
1. Rationality	100.70
2. Curiosity	72.20
3. Open-mindedness	79.70
4. Aversion to Superstitious Beliefs	32.70
5. Objectivity-Intellectual Honesty	63.00
6. Suspended Judgement	34.30
7. Theory Building	13.70
8. On all the dimensions	396.30

If the subject "agreed" with the positive statements and "disagreed" with the negative statements, the scale values that were assigned to the statements were attributed to the individual as positive scores; if the subject "disagreed" with the positive statements and agreed with the negative statements, the scale values that were assigned to the statements were attributed to the individual as negative scores; and if the subject has not answered a particular item, then the zero value was assigned to that item. So in this way, the sums of the total scores of 48 items were calculated for each subject. And then, the data was analysed variable-wise and variable-wise "Means" and "Standard Deviations" of Scientific Attitude scores were computed. Variable-wise the data were subjected to ANOVA (Analysis of Variance) and significant differences were found between various means by computing t-tests.

The correlation of each dimension of the Scientific Attitude with that of the other dimensions and the total dimensions of the Scientifc Attitude were calculated and were arranged in a matrix to know the intercorrelations among the dimensions.

Results and Discussion of Results

As already stated one of the main objectives of the present study was to measure Scientific Attitude among the Academics and Research Scholars in Universities in relation to their demographic factors. Thus the main concern of the present investigation was to find out the relationships between the following pairs of variables.

1. Sex and Scientific Attitude.
2. Occupation and Scientific Attitude.
3. Age and Scientific Attitude.
4. Income and Scientific Attitude.
5. Content Background and Scientific Attitude.
6. Experience and Scientific Attitude.
7. Type of Qualification and Scientific Attitude.
8. Political Ideology and Scientific Attitude.
9. Political Activism and Scientific Attitude.
10. Political Commitment and Scientific Attitude.

This order of the pairs of variables corresponds with the order of the Hypotheses presented under caption 1.8.0. When the data obtained were treated by the Statistical techniques explained under Caption 3.4.0. certain results were arrived at and they are presented in the following pages in order of the hypotheses formulated. After each hypotheses is presented under a caption, the relevant results are furnished with the help of two types of tables.

The first type of tables gives the details of subjects involved variable-wise with their mean Scientific Attitude scores and F-ratio (capital-analysis of variance) between them to underline if the difference is statistically significant or not. The second type of tables give the variable-wise details of the mean differences of Scientific Attitude scores.

In the end, it is stated if the results reject, do not reject or partially reject the hypothesis. Under the sub caption B of each caption, a discussion of the results is presented against the background of the research findings available.

1.A. Hypothesis (1)

There will be no Significant Difference in the Mean Scientific Attitudes of Males and Females.

A null hypothesis of no significant difference in the mean Scientific Attitudes of Males and Females was formulated.

The sample obtained was classified on the basis of sex and the "Means" and "Standard Deviations" of Scientific Attitude scores of Males and Females were computed dimension-wise and total dimensions-wise. Then the data was subjected to t-test to find out if the differences in the mean scores were significant or not. The results obtained in this regard are presented in the Table 3.10.

Table 3.10 reveals that while Males scored a mean score of 41.33 on the Rationality dimension of Scientific Attitude, the

Females scored 37.26. The standard deviations of their scores were 29.88 and 28.91 respectively. The t-value 1.05 obtained between their mean scores was not found to be statistically significant.

On the Curiosity dimension of the Scientific Attitude Scale the Males scored a mean score of 38.71 and the Females scored 36.21 and the standard deviations of their scores were 26.07 and 27.51 respectively. The t-value 0.73 obtained between their mean scores was not found to be statistically significant.

Table 3.10

T-Values between The Mean Scientific Attitude Scores of Males and Females-Dimension-wise and Total Dimensions-wise

Dimension of the Scientific Attitude	*Males (N=441)*	*Females (N=68)*	*t-value*
Rationality (D_1)	M = 41.33 SD = 29.88	M – 37.26 SD = 28.91	1.05
Curiosity (D_2)	M = 38.71 SD = 26.07	M= 36.21 SD = 27.54	0.73
Open Mindedness (D_3)	M = 25.23 SD= 26.22	M = 20.92 SD = 23.47	1.28
Aversion to Superstitious Beliefs (D_4)	M = 14.49 SD = 14.71	M = 13.25 SD = 15.14	0.65
Objectivity-Intellectual Honesty (D_5)	M = 24.94 SD = 21.08	M = 21.21 SD = 23.21	1.34
Suspended Judgement (D_6)	M = 18.64 SD = 15.39	M = 18.91 SD = 14.99	0.14
Theory Building (D_7)	M = 4.06 SD = 7.28	M = 2.16 SD = 4.91	2.09*
Total Dimensions (D_8)	M = 167.40 SD = 90.41	M =149.91 SD= 90.66	1.48

*Significant at 0.05 level.

On the dimension of the Openmindedness, while the Males had a mean score of 25.23, the Females had 20.92 and the standard deviations of their scores were 26.22 and 23.47 respectively. The t-value (1.28) obtained between the mean scores of Males and Females was not found to be statistically significant.

On the dimension of Aversion to Superstitious Beliefs, the Males had a mean score of 14.46, while the Females had 13.25 and standard deviations of their scores were 14.71 and 15.14 respectively. The t-value (0.65) found between the mean scores of Males and Females was not statistically significant.

On the dimension of Objectivity and Intellectually Honesty, the Males scored a mean score of 24.94 while the Females scored a mean score of 21.21. The standard deviations of their scores were 21.08 and 23.21 respectively. And the t-value (1.34) was not found to be statistically significant.

On the dimension of Suspended Judgement, the Males had a mean score of 18.64 and the Females 18.91 and the standard deviations of their scores were 15.39 and 14.99 respectively. The t-value (0.14) found between their mean scores was not found to be statistically significant.

On the "Theory Building" dimension of the Scientific Attitude, the Males scored a mean score of 4.06 while the Females scored a mean score of 2.16 and the standard deviations of their scores were 7.28 and 4.91 respectively. The t-value (2.09) betweent their mean scores was found to be statistically significant at 0.05 level. That is, Males scored significantly higher than the Females on the "Theory Building" dimension of the Scientific Attitude.

On the total dimensions of the Scientific Attitude, the Males scored a mean score of 167.40 and the Females scored a mean score of 149.91 and the standard deviations of their scores were 90.41 and 90.66 respectively. The t-value (1.48) found between the mean total Scientific Attitude scores of Males and Females was not statistically signifcant.

The results reveal that by and large there was no significant difference in the Scientific Attitude' of Males and Females on all the dimensions except in the dimension of Theory Building. And thus the null hypothesis of no significant difference in the mean Scientific Attitudes of Males and Females is not rejected.

2.B. Discussion

While the research under report does not yield any difference in the mean Scientific Attitudes of Male and Female Academics and Research Scholars the researches undertaken in the field do not yield unequivocal trend. Srivastava (1980) studying Scientific Attitude in the Indian context has reported differences in the scores of Boys and Girls on the Scientific Attitude scores. But he did not find any differences in the mean Scientific Attitude scores of Male and Female teachers.

There is a possibility that while the Boys and Girls differ at the early ages they do not differ in Scientific Attitude as they advance in age and particularly when both Men and Women are exposed to Education. The differences in Scientific Attitude scores between the Boys and Girls could be traced to differences in exposure of Boys and Girls to outside world. Girls in the Indian context are mostly confined to the homes and if the parents are not literate and properly educated, it is the tradition that conditions the home environment. Boys could be conceived to have greater exposure to the outside world which may not be very traditional in substance. But if the Boys and Girls are exposed to Modern Knowledge and Culture continuously for longer durations of time it is likely that they do not differ in their Scientific Attitude.

Even in the present study no significant difference in the Male and Female Academics and Researchers was found. The Male and Female Academics and Researchers in the

Universities have been continuously exposed to Modern Knowledge and Culture for longer durations of time. This could be the reason for their not displaying any differences in Scientific Attitudes. The Male and Female Teachers who formed the sample of Srivastava (1980) too have not displayed any differences in the Scientific Attitude scores. So it is not surprising that the Male and Female Academics and Researchers who have been exposed to Modern Knowledge to greater lengths of time have not displayed any difference in their Scientific Attitude in the present study.

Thus it is concluded that Males and Females as such do not differ in their Scientific Attitude.

2.A. Hypothesis (2)

There will be no Significant Difference in the mean Scientific attitudes of Academics and research scholars.

A null hypothesis of no significant difference in the mean Scientific Attitudes of Academics and Research Scholars was formulated.

The sample was classified as Academics and Research Scholars. The "Means" and "Standard Deviations" of their Scientific Attitude scores were computed dimension-wise and total dimensions-wise along with the t-tests to find if the differences in the mean Scientific Attitudes were significant. The results thus obtained in this context are presented in the Table 3.11.

Table 3.11 reveals that while Academics scored a mean score of 43.10 on the Rationality dimension of the Scientific Attitude, the Research Scholars scored 34.68. The standard deviations of their scores were 30.45 and 27.00 respectively. The t-value computed (2.87) between their mean scores was found to be statistically significant at 0.01 level. It was found that Academics scored significantly higher on the dimensions of Rationality of Scientific Attitude Scale than Research Scholars.

Table 3.11

T-Values between the Mean Scientific Attitude Scores of Academics and Research Scholars Dimension-wise and Total Dimensions-wise

Dimension of the Scientific Attitude	*Academics (N= 369)*	*Research Scholars (N=140)*	*t-value*
Rationality (D_1)	M = 43.10 SD = 30.45	M = 34.68 SD = 27.00	2.87**
Curiosity (D_2)	M = 42.99 SD = 24.29	M = 26.21 SD = 27.43	6.71**
Open Mindedness (D_3)	M = 26.03 SD = 26.91	M = 19.47 SD = 22.26	2.80**
Aversion to Superstitious Beliefs (D_4)	M = 15.05 SD = 14.42	M = 12.41 SD = 15.52	1.80
Objectivity-Intellectual Honesty (D_5)	M = 26.23 SD = 20.74	M = 19.73 SD = 22.44	3.09**
Suspended Judgement (D_6)	M = 20.22 SD = 14.79	M = 14.63 SD = 16.03	3.72**
Theory Building (D_7)	M = 4.20 SD = 7.13	M = 2.78 SD = 6.70	2.04*
Total Dimensions (D_8)	M = 178.40 SD = 88.68	M =129.92 SD = 86.19	5.55**

** Significant at 0.01 level
* Significant at 0.05 level

On the Curiosity dimension of the Scientific Attitude the Academics had a mean score of 42.99, while the Research Scholars had 26.21. The standard deviations of their scores were 24.29 and 27.43 respectively. The t-value (6.71) found between the mean scores of Academics and Research Scholars was found to be statistically significant at 0.01 level. Thus the Academics scored significantly higher on the dimension of Curiosity than the Research Scholars.

On the dimension of Open-Mindedness, while the Academics scored a mean score of 26.03, the Research Scholars scored 19.47. The standard deviations of their scores were 26.91 and 22.26 respectively. The t-value (2.80) obtained between their mean scores was statistically significant at 0.01 level. So, it could be said that Academics scored significantly higher on the dimension of Open-Mindedness than the Research Scholars.

On the dimension of Aversion to Superstitious Beliefs, the Academics scored a mean score of 15.05 while the Research Scholars scored a mean score of 12.41. The standard deviations of their scores were 14.42 and 15.52 respectively. The t-value (1.80) found between their mean scores was not found to be statistically significant. Thus it could be said that the Academics and Research Scholars do not differ in scoring on the dimension of Aversion to Superstitious Beliefs.

On the dimension of Objectivity and Intellectual Honesty, the Academics had a mean score of 26.23 and the Research Scholars 19.73 and the standard deviations of their scores were 20.74 and 22.44 respectively. The t-value (3.09) found between their mean scores was statistically significant at 0.01 level. So it could be said that the Academics scored significantly higher on the dimension of Objectivity and Intellectual Honesty than the Research Scholars.

On the dimension of Suspended Judgement, while the Academics scored a mean score of 20.22, the Research Scholars scored 14.63. The standard deviations of their scores were 14.79 and 16.03 respectively. The t-value (3.72) obtained between their mean scores was statistically significant at 0.01 level. Thus the Academics scored significantly higher on the dimension of Suspended Judgement than the Research Scholars.

On the "Theory Building" dimension of the Scientific Attitude, the Academics had a mean score of 4.20 and the

Research Scholars 2.78, and the standard deviations of their scores were 7.13 and 6.70 respectively. The t-value (2.04) found between their mean scores was found to be statistically significant at 0.05 level. Thus the Academics scored significantly higher on the Theory Building dimension of the Scientific Attitude than the Research Scholars.

On the Total dimensions of the Scientific Attitude, the Academics scored a mean score of 178.40 and the Research Scholars scored a mean score of 129.92 and the standard deviations of their scores were 88.68 and 86.19 respectively. The t-value (5.55) found between the mean total Scientific Attitude scores of Academics and Research Scholars was found to be statistically significant at 0.01 level. It was found that Academics scored significantly higher on the total dimensions of the Scientific Attitude Scale than the Research Scholars.

The results reveal that by and large there were significant differences in the Scientific Attitude scores of Academics and Research Scholars on all the dimensions of Scientific Attitude except the dimension of Aversion to Superstitious Beliefs. There was significant difference between the mean scores of Academics and Research Scholars. Thus it could be said that the Academics scored significantly higher than the Research Scholars on all the dimensions and on the total dimensions of Scientific Attitude except the dimension of Aversion to Superstitious Beliefs. Even on the dimension of Superstitious Beliefs the Academics scored more than the Research Scholars. But their superiority was not statistically established. And so the null hypothesis of no significant difference in the mean Scientific Attitudes of Academics and Research Scholars is rejected.

2.B. Discussion

The review of related research literature has not yielded any evidence in this regard. So it can be claimed that the research under report is first of its kind in making a comparative study of the Scientific Attitude of Academics and Research Scholars in the Universities.

The Academics in Universities are the people who have been continuously exposed to Modern Knowledge and Scientific Method of Research for longer durations than the Research Scholars who have been inducted into the University life for the first time and who have just started to understand the Scientific Method of Investigation and its implications. Obviously the Research Scholars scored less on the Scientific Attitude Scale when compared to the Academics who are invariably senior to them. Thus a longer duration of the exposure of the Academics to Modern Knowledge and the methods of Scientific Investigation explains their superior score on the Scientific Attitude Scale.

3.A. Hypothesis (3)

There will be no Significant Difference Between the Age levels of the Subjects and their Mean Levels of Scientific Attitude.

A null hypothesis of no significant difference between the age levels of the subjects and their mean levels of Scientific Attitude was formulated.

The sample obtained was divided for the sake of convenience into four age cohorts, namely 21-30 years, 31-40 years, 41-50 years and 51 years and above. "Means" and "Standard Deviations" of the respective age cohorts were computed and by subjecting the data to ANOVA, F-ratio between the mean scores was found out to underline if the difference in the mean scores was significant or not. The data thus analysed for the purpose is presented in the Table 3.12.

Table 3.12 yields a direct relationship between the age levels and their mean levels of Scientific Attitude.

The F-ratio (10.80) between their mean scores was found to be statistically significant at 0.01 level.

Thus it could be said that age had influenced the Scientific Attitude score.

Table 3.12

F-Ratio between the Mean Scientific Attitude Scores of the Subjects Belonging to Various Age Cohorts

Age Cohorts	*Strength (N)*	*Mean*	*S.D.*	*F-ratio*
21-30 years	140	133.47	85.64	
31-40 years	140	170.16	92.04	F= 10.80**
41-50 years	147	174.25	88.30	
51 years and above	76	199.96	80.61	

** Significant at 0.01 level

The Table 3.13 reveals the differences in the mean Scientific Attitude scores of the subjects belonging to various age levels.

Table 3.13

Differences in the Mean Scientific Attitude Scores of the Subjects belonging to Various Age Levels

	31-40 yrs.	*41-50 yrs.*	*51 yrs. & above*
21-30 yrs	36.69** (31-40 yrs)+	40.78** (41-50 yrs)+	66.49** (51 yrs & above)+
31-40 yrs		4.09 (41-50 yrs)+	29.8* (51 yrs & above)+
41-50 yrs			25.71* (51 yrs & above)+

** Significant at 0.01 level.

* Significant at 0.05 level.

( )+ The group shown in parantheses with plus mark got the higher mean score than the other group under comparison.

Table 3.13 reveals the following:

1(a) Subjects belonging to (21-30) age cohort scored the least on the Scientific Attitude Scale. They scored less than the subjects belonging to the rest of the age cohorts i.e., (31-40), (41-50) and (51 years and above).

(b) The difference in the mean Scientific Attitude scores among the groups was found to be statistically significant at 0.01 level.

(c) The differences in the mean scores increased with their age levels.

2 (a) Subjects belonging to (31-40) age cohort scored lesser on the Scientific Attitude Scale than the subjects belonging to age cohorts, (41-50) and (51 years and above).

(b) The difference in the mean Scientific Attitude scores of (31-40) and (41-50) age cohorts was not found to be statistically significant.

(c) The difference in the mean Scientific Attitude scores of cohorts (31-40) and (51 years and above) was found to be statistically significant at 0.01 level.

3(a) Subjects belonging to (41-50) age cohort scored significantly less on the Scientific Attitude Scale than the subjects belonging to (51 years and above) age cohort as mean differences between them was found to be statistically significant at 0.01 level.

The results presented in the Tables 3.12 and 3.13 reveal that the age levels of the subjects have association with their Scientific Attitude scores. The higher the age level, the higher was the Scientific Attitude score. Thus it could be established that the age has positive association with the Scientific Attitude score. And so the null hypothesis of no significant difference between the age levels of the subjects and their mean levels or Scientific Attitude gets rejected.

3.B Discussion

The review of research presented in Chapter-II has not thrown any light on the relationship between Age and Scientific Attitude.

The discussions presented under captions 1.B. and 2.B. indirectly provide some base for the discussion on the relationship between Age and Scientific Attitude. It has been observed that while Boys and Girls differed in their Scientific Attitude scores, the Male and Female teachers did not, and while the Research Scholars scored less on the Scientific Attitude Scale, the Academics, who were relatively seniors, have scored significantly higher. One common element that underscores the discussions presented is that with seniority and Age the Scientific Attitude increases particularly when subjects are continuously exposed to Modern Knowledge and Methods of Scientific Investigation.

Even in the present study the same trend gets focussed, that is, the scores of Academics and Research Scholars increase with Age. The implication is the same. That is, the Academics and Research Scholars whose very business is to acquire more Modern Knowledge and to pursue Scientific Research increase their scores on Scientific Attitude Scale with Age.

4.A Hypothesis (4)

There will be no significant difference in the mean levels of scientific attitude of Academics and Research scholars because of differences in their income levels.

A null hypothesis of no significant difference in the mean levels of Scientific Attitudes of Academics and Research Scholars because of the differences in their income levels was formulated.

The sample thus divided on the basis of six income levels, is as follows:

(i) Rs. 500/- and below per month
(ii) Rs. 500 - 1000/- per month
(iii) Rs.1000 - 2500/- per month
(iv) Rs.2500 - 5000/- per month
(v) Rs.5000 - 8000/- per month
(vi) Rs.8000 and above per month

For the sake of statistical convenience the income levels of Rs. 500/- and below and Rs.500-1000/- per month were clubbed together and named it as Rs.1000/- and below. The "Means" and "Standard deviations" of the Scientific Attitude scores of the subjects belonging to five income levels were computed. The data was subjected to Analysis of Variance to find if there could be any differences between the five sets of mean scores. The data thus analysed is presented in the Table 3.14.

Table 3.14

F-Ratio between the Mean Scientific Attitude Scores of the Subjects Belonging to Various Income Levels

Income levels	*Strength (N)*	*Mean*	*S.D.*	*F-ratio*
Rs.1000 and below	56	99.69	85.71	
Rs. 1000 - 2500	72	138.17	82.09	
Rs. 2500 - 5000	189	176.76	91.56	F = 13.40**
Rs. 5000 - 8000	167	186.09	83.59	
Rs. 8000 and above	8	200.04	66.52	

** Significant at 0.01 level.

Table 3.14 reveals that the mean Scientific Attitude scores of the subjects increased with their income levels and the mean differences among them was found to be statistically significant at 0.01 level. And so to find out the mean differences among the five sets of income levels the t-ratios were found out. The results thus found out are presented in the Table 3.15.

Table 3.15

Differences in the Mean Scientific Attitude Scores of the Subjects Belonging to Various Income Levels

	Rs. 1000-2500	*Rs. 2500-5000*	*Rs.5000-8000*	*Rs.8000 and above*
Rs. 1000 and below	38.48* (1000-2500)+	77.07** (2500-5000)+	86.40** (5000-8000)+	100.35** (8000 & above)+
Rs. 1000-2500		38.59** (2500-5000)+	47.92** (5000-8000)+	66.27** (8000 & above)
Rs. 2500-5000			9.33 (5000-8000)+	23.28 (8000 & above)+
Rs. 5000-8000				13.95 (8000 & above)+

** Significant at 0.01 level
**Significant at 0.05 level
( )+ The group shown in paranthesis with plus marks got higher mean score than the other, group under comparison.

Table 3.15 reveals that:

1. (a) The subjects belonging to Rs.1000 and below income level scored the least on Scientific Attitude Scale.

   (b) The difference in the mean Scientific Attitude scores of the subjects belonging to Rs.1000 and below and Rs.1000-2500 was found to be statistically significant at 0.05 level. The differences in the mean Scientific Attitude scores of the subjects belonging to Rs.1000 and below and that of the subjects belonging to 2500-5000, Rs.5000-8000 and above were found to be statistically significant at 0.01 level.

(c) Subjects belonging to Rs.1000 and below income level scored significantly less on the Scientific Attitude Scale than the subjects belonging to the other four income levels.

(d) The difference in the mean Scientific Attitude scores of the subject increased with their income levels.

2. (a) Subjects belonging to the income level of Rs.1000-2500 scored lesser on the Scientific Attitude Scale than the subjects belonging to the income levels of Rs. 2500-5000, Rs. 5000-8000 and Rs. 8000 and above. The mean differences among the three groups were found to be statistically significant at 0.01 level.

(b) Subjects belonging to the income level of Rs.1000-2500 significantly scored less than the subjects belonging to the income levels of Rs. 2500-5000, Rs. 5000-8000 and Rs. 8000 and above.

(c) The difference in the mean Scientific Attitude scores of the subjects increased with their income levels.

3. (a) Subjects belonging to the income level of Rs. 2500-5000 scored less on the Scientific Attitude Scale than the subjects belonging to the income levels of Rs. 5000-8000 and as Rs. 8000 and above.

(b) Subjects belonging to the income levels of Rs. 2500-5000 scored less on the Scientific. Attitude Scale than the subjects belonging to income levels of Rs. 5000-8000 and Rs. 8000 and above. But the mean difference among the two groups were not found to be statistically significant.

4. Subjects belonging to the income level of Rs. 5000-8000 scored less on the Scientific Attitude Scale than the subjects belonging to the income level of Rs. 8000 and above. But the difference between their mean scores was not found to be statistically significant.

The results presented in the Table 3.14 and Table 3.15 yield a positive relationship between the income levels of the subjects and their levels of Scientific Attitude. Thus it could be established that the income levels of the subjects have a positive association with their Scientific Attitude scores. So the null hypothesis of no significant difference in the mean levels of Scientific Attitude of Academics and Research Scholars because of the differences in their income levels is rejected.

4.B. Discussion

The review of research presented does not include any researches which underline the relationship between the Income levels of any groups of subjects and their Scientific Attitude scores. The present research yields a positive association between the Income levels and the scores on Scientific Attitude Scale. But it must be mentioned that the positive association found between Income levels and the scores on Scientific Attitude Scale can not be stretched beyond the present sample of Academics and Research Scholars.

The Academics improve their Income levels with both seniority and the acquisition of new knowledge. The Income level increase expeditiously when the Academics conduct more researches and acquire the research degree for themselves and for their own students. The increase in the Scientific Attitude scores with more Income could be an exclusive quality of the Academics. And thus the explanation for the positive association between Income levels and the levels of Scientific Attitudes of Academics and Research Scholars.

5.A. Hypothesis (5)

There will be no Significant Difference in the Mean Levels of Scientific Attitude of Academics and Research Scholars because of the Differences in their Content background.

A null hypothesis of no significant difference in the mean levels of Scientific Attitude of the subjects because of the differences in their content background was formulated.

The sample was classified on the basis of the main content background of the subjects and their means and standard deviations on Scientific Attitude scores were computed. Then the data was subjected to the Analysis of Variance to find out F-ratio to underline if the difference among the mean scores was significant or not. The results obtained through the statistical treatment are presented in the Table 3.16.

Table 3.16

F-Ratio between the Mean Scientific Attitude Scores of the Subjects Belonging to Various Content Backgrounds

S. No.	*Content Background of the subject*	*Strength (N)*	*Mean*	*S.D.*	*F-ratio*
1.	Humanities & Arts Subjects	71	154.28	90.95	
2.	Social Science Subjects	95	171.63	90.91	
3.	Physical Science Subjects	103	186.29	88.59	
4.	Biological Science Subjects	66	129.96	76.46	F=3.22**
5.	Health Science	86	158.85	93.32	
6.	Engineering	58	169.80	98.94	
7.	Mathematics	30	182.85	78.02	

** Significant at 0.01 level.

Table 3.16 reveals that the subjects belonging to Physical Sciences scored the highest on the Scientific Attitude Scale, while the subjects belonging to Biological Sciences scored the least. The order of the mean scores of other groups obtained is as follows:

1. Physical Sciences (P.Sc.) 186.29
2. Mathematics (M) 182.85
3. Social Sciences (S.Sc)) 171.63

4. Engineering (Engg.) 169.80
5. Health Sciences (H.Sc.) 158.85
6. Humanities and Arts (H & A) 154.28
9. Biological Sciences (B.Sc.) 129.96

The F-ratio (3.22) computed between the mean scores of seven groups was found to be statistically significant at 0.01 level. That is to say that the Content background of the subjects has an association with their Scientific Attitude.

To find out the mean differences in the Scientific Attitude scores of each of the group when compared with other groups, the t-tests were employed. The results thus obtained are presented in the Table 3.17.

Table 3.17

Differences in the Mean Scientific Attitude Scores of the Subjects Belonging to Various Content Backgrounds

	S.Sc.	*P.Sc.*	*B.Sc.*	*H.Sc.*	*Engg.*	*Maths.*
Humanities and Arts	17.35 (S.Sc)+	32.01* (P.Sc)+	24.32 (H&A)+	4.37 (H.Sc)+	15.52 (Engg)+	28.57 (Maths)+
Social Sciences		14.66 (P.Sc)+	41.67** (S.Sc)+	12.78 (S.Sc)+	1.83 (S.Sc)+	11.22 (Maths)+
Physical Sciences			56.33** (P.Sc)+	27.44* (P.Sc)+	16.49 (P.Sc)+	3.44 (P.Sc)+
Biological Sciences				28.89 (H.Sc)+	39.8* (Engg)+	5.89** (Maths)+
Health Sciences					10.95 (Engg.)+	24.00 (Maths)+
Engineering						13.05 (Maths)+

** Significant at 0.01 level.

* Significant at 0.05 level.

( )+ The group shown in parantheses with plus mark got the higher mean score than the other group under comparison.

1. (a) Subjects belonging to Humanities and Arts group scored less on the Scientific Attitude Scale than the subjects belonging to Social Sciences, Physical Sciences, Health Sciences, Engineering and Mathematics. The differences among their mean scores, however, were not found to be statistically significant barring the physical sciences group.

   (b) Subjects belonging to Humanities and Arts group significantly scored less on the Scientific Attitude Scale than the subjects belonging to the Physical Sciences group. The difference in their mean scores was statistically significant at 0.05 level.

   (c) Subjects belonging to Humanities and Arts group scored higher on the Scientific Attitude Scale than the subjects belonging to Biological Sciences. The difference, however in their mean scores was not found to be statistically significant

2. (a) Subject belonging to Social Sciences scored less on the Scientific Attitude Scale than the subjects belonging to Physical Sciences and Mathematics. The differences in their mean scores however were not statistically significant.

   (b) Subjects belonging to Social Sciences scored higher on the Scientific Attitude Scale than the subjects belonging to Biological Sciences, Health Sciences and Engineering. But the differences in their mean scores were not found to be statistically significant except in the case of Biological Sciences group.

   (c) Subjects belonging to Social Sciences scored significantly higher on the Scientific Attitude Scale than the subjects belonging to Biological Sciences.

3. (a) Subjects belonging to Physical Sciences scored the highest on the Scientific Attitude Scale.

(b) Subjects belonging to Physical Sciences scored significantly higher on the Scientific Attitude Scale than the subjects belonging to Biological Sciences and Health Sciences as their mean differences were significant at 0.01 and 0.05 levels respectively.

(c) Subjects belonging to Physical Sciences scored higher than the subjects belonging to Engineering and Mathematics. But the difference in their mean scores were not found to be statistically significant.

4. (a) The subjects belonging to Biological Sciences scored the least on the Scientific Attitude Scale.

(b) The subjects belonging to the Biological Sciences significantly scored less on the Scientific Attitude Scale than the subjects belonging to Engineering and Mathematics groups. The mean differences among them were found to be statistically significant at 0.05 and 0.01 levels respectively.

(c) The subjects belonging to Biological Sciences scored less on the Scientific Attitude Scale than the subjects belonging to Health Sciences. But the mean difference was not found to be statistically significant.

5. Subjects belonging to Health Sciences scored less on the Scientific Attitude Scale than the subjects belonging to Engineering and Mathematics groups. But the difference in their mean scores were not statistically significant.

6. Subjects belonging to Engineering scored less on the Scientific Attitude Scale than the subejcts belonging to Mathematics group. But the mean difference was not statistically significant.

The results thus arrived at precisely are:

1. Subjects belonging to the Physical Sciences scored the highest on the Scientific Attitude Scale.

2. Subjects belonging to Physical Sciences scored significantly higher on the Scientific Attitude Scale than the subjects belonging to Humanities, Biological Sciences and Health Sciences.

3. Subjects belonging to Social Sciences scored significantly higher on the Scientific Attitude Scale than the subjects belonging to Biological Sciences.

4. Subjects belonging to Engineering and Mathematics groups scored significantly higher on the Scientific Attitude Scale than the subjects belonging to Biological Sciences.

5. Subjects belonging to Biological Sciences scored the least on the Scientific Attitude Scale. Thus the null hypothesis of no significant difference in the mean levels of Scientific Attitude of Academics and Research Scholars because of the differences in their content background gets rejected.

Further, the present investigator was equally interested in finding out if there could be any significant differences between the mean scores of the subjects scored on each of the dimension of the Scientific Attitude Scale.

So, the data was classified dimension-wise and then it was subjected to Analysis of Variance. The dimension-wise F-ratios found out in this regard are presented sequentially in the following paragraphs through tables containing relevant results under subcaptions. Whenever the F-ratios were found to be significant, the t-tests were computed between the various pairs of means and the results thus obtained are presented in the Tables. The pairs of means which were statistically significant are reported in the following paragraphs and the pairs of means which were not statistically significant are not included in the report.

5.A.1. Rationality

The results pertaining to the Rationality dimension are presented in the Table 3.17.1

Table 3.17.1

F-Ratio between the Mean Scientific Attitude Scores of the Subject on the Dimension of Rationality in Relation to their Content Background

S. No.	*Main content background of the subjects*	*Strength (N)*	*Mean*	*S.D.*	*F-ratio*
1.	Humanities and Arts	71	37.33	32.32	
2.	Social Sciences	95	43.76	30.68	
3.	Physical Sciences	103	41.53	31.95	
4.	Biolgoical Sciences	66	37.28	25.35	F=0.80
5.	Health Sciences	86	43.83	28.16	
6.	Engineering	58	37.16	31.49	
7.	Mathematics	30	42.91	21.99	

Table 3.17.1 reveals that the subjects belonging to Health Sciences scored the highest on the Rationality dimension with a mean score of 43.83 while the subjects belonging to Engineering scored the least with a mean score of 37.16. The mean scores of the other subjects fell in between the range, 37.16-43.83. The F-ratio (0.80) computed between the mean scores was not found to be statistically significant.

Thus, it gets established that on the dimension of Rationality there was no significant differences between the mean scores of the subjects belonging to various content backgrounds.

5.A.2. *Curiosity*

The results pertaining to the dimension of Curiosity are presented in the Table 3.17.2.

Table 3.17.2

F-Ratio between the Mean Scientific Attitude Scores of the Subjects on the Dimension of Curiosity in Relation to their Content Background

S. No.	*Main content background of the subjects*	*Strength (N)*	*Mean*	*S.D.*	*F-ratio*
1.	Humanities and Arts	71	37.57	24.75	
2.	Social Sciences	95	38.25	28.53	
3.	Physical Sciences	103	44.25	24.50	
4.	Biological Sciences	66	27.45	27.60	F=3.88**
5.	Health Sciences	86	36.13	25.01	
6.	Engineering	58	44.27	26.00	
7.	Mathematics	30	44.31	21.03	

** Significant at 0.01 level.

Table 3.17.2 reveals that the subjects belonging to Mathematics background scored the highest on the dimension of Curiosity with a mean score of 44.31 while the subjects belonging to the Biological Sciences background scored the least with a mean score of 27.45. The mean scores of other subjects ranged between 27.45 and 44.31 and the difference in the mean scores was found to be statistically significant at 0.01 level. Thus it could be said that the content background of the subjects influenced significantly the scores of the subjects on the dimension of Curiosity.

The differences in the pairs of mean scores of the subjects on the dimension of Curiosity was tested through t-test. The results thus obtained are presented in the table 3.17.3.

Table 3.17.3

Differences in the Mean Scientific Attitude Scores of the Subjects of the Dimension of Curiosity in Relation to their Content Background

	S.Sc.	*P.Sc.*	*B.Sc.*	*H.Sc*	*Engg.*	*Maths.*
Humanities and Arts	2.68 (S.Sc)+	8.68* (P.Sc)+	8.12 (H&A)+	0.56 (H.Sc)+	8.70 (Engg)+	8.74 (Maths)+
Social Sciences		6.00 (P.Sc)+	10.80 (S.Sc)+	2.12 (S.Sc)+	6.02 (Engg)+	6.06 (Maths)+
Physical Sciences			16.80** (P.Sc)+	8.12* (P.Sc)+	0.02 (Engg)+	0.06 (Maths)+
Biological Sciences				8.68* (H.Sc)+	16.82** (Engg)+	16.86** (Maths)+
Health Sciences					8.14 (Engg)+	8.18 (Maths)+
Engineering						0.04 (Maths)+

** Significant at 0.01 level

* Significant at 0.05 level

( )+ The group shown in the parantheses with plus mark got the higher mean score than the other group under comparison.

Table 3.17.3 reveals that:

1. The subjects belonging to Physical Sciences scored significantly higher on the dimension of Curiosity than the subjects belonging to Humanities, Biological Sciences and Health Sciences as their differences in the mean scores were statistically significant.

2. Subjects belonging to Health Sciences scored significantly higher on the Curiosity dimension than the subjects belonging to Biological Sciences. The difference between their mean scores was statistically significant at 0.01 level.

3. Subjects belonging to Mathematics and Engineering groups scored significantly higher on the dimension of Curiosity than the subjects belonging to Biological Sciences.

Thus it could be established that the subjects belonging to Physical Sciences, Engineering, Mathematics and Health Sciences were found to be more Curious than subjects from all other subject groups. Subjects belonging to the Biological Sciences were found to be less Curious than other subjects.

5.A.3. Open-Mindedness

The results pertaining to the dimension of Open-mindedness are presented in the Table 3.17.4 and 3.17.5.

Table 3.17.4

F-Ratio between the Mean Scientific Attitude of Subjects on the Dimension of Open-Mindedness in Relation to their Content Background

S. No.	*Main content background of the subjects*	*Strength (N)*	*Mean*	*S.D.*	*F-ratio*
1.	Humanities and Arts	71	29.94	23.59	
2.	Social Sciences	95	25.74	26.94	
3.	Physical Sciences	103	30.13	24.90	
4.	Biological Sciences	66	14.78	23.21	F=5.40**
5.	Health Sciences	86	15.65	24.98	
6.	Engineering	58	27.52	25.92	
7.	Mathematics	30	31.90	28.27	

** Significant at 0.01 level

Table 3.17.4 reveals that while the subjects belonging to Mathematics scored the highest on the dimension of Open-Mindedness with a mean score of 31.90, the subjects belonging to Biological Sciences scored the least with a mean score of 14.78. The other means varied in the range of 17.12. The F-ratio (5.40) obtained between their mean scores was found to be statistically significant at 0.01 level. Thus it could be said that the content background of the subjects influenced the Scientific Attitude score on the dimension of Open-Mindedness.

The differences that were observed between the mean scores were tested through t-tests. The results thus obtained are presented in the Table 3.17.5.

Table 3.17.5

Differences in the Mean Scientific Attitude Scores of the Subjects on the Dimension of Openmindedness in Relation to their Content Background

	S.Sc.	*P.Sc.*	*B.Sc.*	*H.Sc*	*Engg.*	*Maths.*
Humanities and Arts	4.20 (H&A)+	0.19 (P.Sc)+	15.16** (H&A)+	14.29** (H&A)+	2.42 (H&A)+	1.96 (Maths)+
Social Sciences		4.39 (P.Sc)+	10.96** (S.Sc.)+	10.09** (S.Sc)+	1.78 (Engg)+	6.16 (Maths)+
Physical Sciences			15.35** (P.Sc)+	14.48** (P.Sc)+	2.61 (P.Sc)+	1.77 (P.Sc)+
Biological Sciences				0.87 (H.Sc)+	12.74** (Engg)+	17.12** (Maths)+
Health Sciences					11.87** (Engg)+	16.25** (Maths)+
Engineering						1.38 (Maths)+

** Significant at 0.01 level
* Significant at 0.05 level
( )+ The group shown in parantheses with plus mark got the higher mean score than the other group under comparison.

Table 3.17.5 reveals that:

1. The subjects belonging to Humanities and Arts background scored significantly higher on the dimension of Open-mindedness than the subjects belonging to the Biological Sciences and Health Sciences as their differences were significant at 0.01 levels;

2. the subjects belonging to the Social Sciences scored significantly higher on the dimension of Open-mindedness than the subjects belonging to the Biological Sciences and Health Sciences. The mean differences among the groups were statistically significant at 0.01 level;
3. the subjects belonging to Physical Sciences scored significantly higher on the dimension of Open-mindedness than the subjects belonging to the Biological Sciences and Health Sciences. The differences in their mean scores were found to be statistically significant at 0.01 level; and
4. the subjects belonging to Engineering and Mathematics groups scored significantly higher on the dimension of Open-mindedness than the subjects belonging to the Biological Sciences and Health Sciences as their mean differences were found to be statistically significant at 0.01 level.

5.A.4. Aversions to Superstitious Beliefs

The results pertaining to the dimension of the Aversion to Superstitious Beliefs are presented in the Tables 3.17.6 and 3.17.7.

Table 3.17.6 reveals that the subjects belonging to Social Sciences scored the highest on the dimension of Aversion to Superstitious Beliefs with a mean score of 16.65 while the subjects belonging to Biological Sciences scored the least with a mean score of 8.69. Thus the other means varied within the range of 7.96. The F-ratio (2.20) obtained between their mean scores was found to be statistically significant at 0.05 level. And so it could be said that the content background of the subjects influenced the Scientific Attitude score on the dimension of Aversion to Superstitious Beliefs.

Table 3.17.6

F-Ratio Between the Mean Scientific Attitude Scores of the Subjects on the Dimension of Aversian to Superstitious Beliefs in Relation to their Content Background

S. No.	*Main content background of the subjects*	*Strength (N)*	*Mean*	*S.D.*	*F-ratio*
1.	Humanities and Arts	71	14.79	15.22	
2.	Social Sciences	95	16.65	12.19	
3.	Physical Sciences	103	15.61	13.56	
4.	Biological Sciences	66	8.69	17.49	F=2.20**
5.	Health Sciences	86	14.70	14.67	
6.	Engineering	58	13.56	14.96	
7.	Mathematics	30	14.25	16.58	

* Significant at 0.05 level.

The differences that were observed in the mean scores were tested through t-tests and the results obtained are presented in the Table 3.17.7

Table 3.17.7

Differences in the Mean Scientific Attitude Scores of the Subjects on the Dimension of Aversion to Superstitious Beliefs in Relation to their Content Background

	S.Sc.	*P.Sc.*	*B.Sc.*	*H.Sc.*	*Engg.*	*Maths.*
Humanities and Arts	1.86 (S.Sc)+	0.82 (P.Sc)+	6.10** (H&A)+	0.09 (H&A)+	1.23 (H&A)+	0.54 (H&A)+
Social Sciences		1.04 (S.Sc)+	7.96** (S.Sc)+	1.95 (S.Sc)+	3.09 (S.Sc)+	2.40 (S.Sc)+
Physical Sciences			6.92* (P.Sc)+	0.91 (P.Sc)+	2.05 (P.Sc)+	1.36 (P.Sc)+
Biological Sciences				6.01* (H.Sc)+	4.87 (Engg)+	5.56 (Maths)+
Health Sciences					1.14 (H.Sc)+	0.45 (H.Sc)+
Engineering						0.69 (Maths)+

** Significant at 0.01 level. * Significant at 0.05 level.

( )+ The group shown in parantheses with plus mark got the higher mean score than the other group under comparison.

Table 3.17.7. reveals that:

The subjects belonging to Humanities and Arts, Social Sciences, Physical Sciences and Health Sciences scored significantly higher on the dimension of Aversion to Superstitious Beliefs than the subjects belonging to Biological Sciences. The mean differences among them were found to be statistically significant.

5.A.5. *Objectivity and Intellectual Honesty*

The results pertaining to the dimension of Objectivity and Intellectual Honesty are presented in the Tables 3.17.8 and 3.17.9.

Table 3.17.8

F-Ratio between the Mean Scientific Attitude Scores of the Subjects on the Dimension of Objectivity and Intellectual Honesty in Relation to their Content Background

S. No.	*Main content background of the subjects*	*Strength (N)*	*Mean*	*S.D.*	*F-ratio*
1.	Humanities and Arts	71	18.58	20.12	
2.	Social Sciences	95	23.30	22.29	
3.	Physical Sciences	103	31.27	19.54	
4.	Biological Sciences	66	22.08	21.30	F=2.89**
5.	Health Sciences	86	24.59	22.63	
6.	Engineering	58	24.10	22.07	
7.	Mathematics	30	23.93	18.57	

** Signifcant at 0.01 level.

Table 3.17.8 reveals that while the subjects belonging to the Physical Sciences scored the highest on the dimension of Objectivity and Intellectual Honesty with a mean score of 31.27, the subjects belonging to the Humanities and Arts group scored the least with a mean score of 18.58. The other means varied within the range 18.58 and 31.27. The F-ratio (2.89) obtained between their mean scores was found to be statistically significant at 0.01 level. Thus, it could be said that the content background of the subjects influenced the Scientific Attitude score on the dimension of Objectivity and Intellectual Honesty.

The differences in the mean scores were tested through t-tests, and the results thus obtained are presented in the Table 3.17.9.

Table 3.17.9

Differences in the Mean Scientific Attitude Scores of the Subjects on the Dimension of Objectivity and Intellectual Honesty in Relation to their Content Background

	S.Sc.	*P.Sc.*	*B.Sc.*	*H.Sc.*	*Engg.*	*Maths*
Humanities and Arts	4.72 (S.Sc)+	12.69** (P.Sc)+	3.50 (B.Sc)+	6.01 (H.Sc)+	5.52 (Engg)+	5.35 (Maths)+
Social Sciences		7.97** (P.Sc)+	1.22 (S.Sc)+	1.29 (H.Sc)+	0.80 (Engg)+	0.63 (Maths)+
Physical Sciences			9.19* (P.Sc)+	6.68* (P.Sc)+	7.17* (P.Sc)+	7.34 (P.Sc)+
Biological Sciences				2.51 (H.Sc)+	2.02 (Engg)+	1.85 (Maths)+
Health Sciences					0.49 (H.Sc)+	0.66 (H.Sc)+
Engineering						0.17 (Engg)+

** Significant at 0.01 level.
* Significant at 0.05 level.
( )+ the group shown in parantheses with plus mark got the higher mean score than the other group under comparison.

Table 3.17.9 reveals that the subjects belonging to The Physical Sciences scored significantly higher on the dimension of Objectivity and Intellectual Honesty than the subjects belonging to Engineering, Health Sciences, Biological Sciences, Social Sciences and Humanities and Arts groups. The mean differences among them were found to be statistically significant. Thus it gets established that the subjects belonging to Physical Sciences are found to have more Objectivity and Intellectual Honesty than the susbjects of all other groups barring the Mathematics group.

5.A.6. Suspended Judgement

The results pertaining to the dimension of Suspended Judgement are presented in the Table 3.17.10.

Table 3.17.10

F-Ratio between the Mean Scientific Attitude Scores of the Subjects of the Dimension of Suspended Judgement in Relation to their Content Background.

S. No.	*Main content background of the subjects*	*Strength (N)*	*Mean*	*S.D.*	*F-ratio*
1.	Humanities and Arts	71	14.50	17.44	
2.	Social Sciences	95	19.50	14.73	
3.	Physical Sciences	103	20.56	15.09	
4.	Biological Sciences	66	16.62	13.27	F=1.96
5.	Health Sciences	86	19.53	15.61	
6.	Engineering	58	17.73	16.36	
7.	Mathematics	30	23.42	12.17	

Table 3.17.10 reveals that the subjects belonging to Mathematics scored the highest on the dimension of Suspended Judgement with a mean score of 23.42, while the subjects belonging to Humanities and Arts group scored the least with a mean score of 14.50 with the mean difference of 8.92. The other means ranged between 14.50 and 23.42. The F-ratio (1.96) obtained between their mean scores was not found to be statistically significant. Thus, it could be established that the content background of the subjects did not influence

the Scientific Attitude score of the subjects on the dimension of Suspended Judgement.

5.A.7. Theory Building

The results pertaining to the dimension, Theory Building, are presented in the Table 3.17.11.

Table 3.17.11

F-Ratio between the Mean Scientific Attitude Scores of the Subjects on the Dimension of Theory Building in Relation to their Content Background

S. No.	*Main content background of the subjects*	*Strength (N)*	*Mean*	*S.D.*	*F-ratio*
1.	Humanities and Arts	71	3.57	8.08	
2.	Social Sciences	95	4.44	6.85	
3.	Physical Sciences	103	2.93	6.90	
4.	Biological Sciences	66	3.05	7.50	F=1.46
5.	Health Sciences	86	4.42	5.45	
6.	Engineering	58	5.45	7.08	
7.	Mathematics	30	2.14	7.09	

Table 3.17.11. reveals that while the subjects belonging to Engineering group scored the highest on the dimension of Theory Building with a mean score of 5.45, the subjects of Mathematics group scored the least with a mean score of 2.14. The mean scores ranged between 2.14 and 5.45. The F-ratio (1.46) obtained between their mean scores was not found to be statistically significant. Thus it gets established that the Scientific Attitude scores on the Theory Building dimension do not differ because of their content background.

The results presented under the captions 3.5.A, 3.5.A.1., 3.5.A.2., 3.5.A.3., 3.5.A.4., 3.5.A.5., 3.5.A.6. and 3.5.A.7 reveal that

1. The content background of the subjects influenced the Scientific Attitude scores of the subjects on the

dimensions of Curiosity, Open-mindedness. Aversion to Superstitious Beliefs and Objectivity and Intellectual Honesty.

2. The content background of the subjects does not influence the Scientific Attitude scores of the subjects on the dimension of Rationality, Suspended Judgement and Theory Building.

3. The content background of the subjects influenced the total Scientific Attitude scored of the subjects on all the dimensions.

5.B. Discussion

A perusal of the results presented in the preceding pages underscores the fact that the sample drawn with the background of multiple academic disciplines and professional areas reflect differences in the mean levels of the scores on the Scientific Attitude Scale. It may be noted that conglomerations of discipline based samples include subjects of all age groups, income groups, and are of both the sexes. So the differences displayed on Scientific Attitude is a reflection of the impact of their content of each discipline. Thus it is concluded that the nature of the content could have a bearing on the Scientific Attitude scores of the Academics and Research Scholars. On the basis of the scores obtained by the groups of subjects they could be put in a continuum ranging from the highest scores obtained to the least scores obtained on the Scientific Attitude Scale. A continuum thus obtained is as follows:

P.Sc.	Maths.	S.Sc.	Engg.	H.Sc.	H&A	B.Sc.
(186.29)	(182.85)	(171.63)	(169.80)	(158.85)	(154.28)	(129.96)

The pattern thus obtained in this research roughly corresponds with the pattern obtained by various researchers all over the World. The researchers that have studied the ideological orientation of the Academics on a Conservative/

Liberal Continuum have found a similar pattern with slight variations (Ladd and Lipset, 1975). Social Scientists were found to be at the top of the Liberal attitude with the Natural Scientists, Engineers and the Agriculture Specialists at the other end of continuum. A question may be posed as to why the Academics with different content background should betray differences on a Liberal/Conservative continuum or on Scientific Attitude Scale? Has the content background anything to do with the Attitude formation? Ladd and Lipset (1975) aver that "The subject defines the problems which occupy their work life. It carries with it a distinctive way of approaching matters, a conceptual frame which is developed in response to the materials and concerns of the discipline, but which is obviously applied beyond the boundaries of the discipline". Ladd and Lipset agree that the subject matter of the field also conditions the associations and identifications of Academics with the out side world of the University. A professor of Engineering typically has lot more contact with the businessman. A faculty member in a college of Agriculture has greater rapport with farmers. Thus the academic content not only provides a distinctive conceptual frame, and way of approach but the facility of rapport with the disparate outside groups. All these together fashion the distinctive styles and modes of thought of Academics resulting in disparate sub-cultures (Ibid:1975). It has been established that certain subject areas foster Liberalism while certain other Conservatism. And while some develop Scientific Attitudes more others do not. Why should it happen? In what way a particular discipline like Physics becomes instrumental in fostering greater measure of the Scientific Attitude and how do content areas like Botany or Zoology or Engineering and Health Sciences restrict such development?

It is argued (Meier, 1951, PP 170) that a discipline like "Physics orients its practitioners in the proposition that progress is made by discarding various assumptions and premises and thereby making it possible to create a more powerful theory upon a simpler underpinning. The Physicist, more than any Scientist, deals with abstractions which make nonsense out of observations based upon the common place; he is educated in doubt and can disregard evidence which is to the ordinary observer is both convincing and conclusive".

Thus it can be concluded that a discipline with wider ramifications wherein there is much scope for abstraction and questioning of assumptions and of building new theories and laws consistently there is a greater scope for the emergence of Liberalism and the development of Scientific Attitude. When the area of operation is a limited specialisation involving repetitive academic exercises the scope for "Doubt" is limited and the practitioners of disciplines tend to become Conservatives and their Conservatism gets reinforced by the nature of the clientele they get because of their peculiar position like Engineers getting the businessmen and the Agricultural Scientist getting the farmers.

Thus Physicists are found at the top on the scores on Scientific Attitude Scale followed by Mathematicians, Social Scientists, Engineers, Health Scientists, and Academics working in the faculties of Arts and Humanities and finally the Botanists and Zoologists.

The only deviation the present research has exemplified is that of the faculty members working in Humanities and Arts taking a back seat in scoring on the Scientific Attitude Scale. This could be mainly because the Academics from Indian Languages predominated the sample chosen and there was a negligible representation of Academics from the discipline of Philosophy, Fine Arts. etc.

6.A. Hypothesis (6)

There will be no Significant Difference Between the Duration of the Subjects' Experience and their mean levels of Scientific Attitude.

A null hypothesis of no significant difference between the duration of the subjects' experience and their mean levels of Scientific Attitude was formulated.

The sample was divided according to the seven experience levels of the subjects. The experience levels thus formed for the purpose of present study are as follows:

(i) 1-5 years experience
(ii) 6-10 years experience
(iii) 11-15 years experience
(iv 16-20 years experience
(v) 21-25 years experience
(vi) 26-30 years experience
(vii) 31-35 years and above

The means and standard deviations of the Scientific Attitude scores of the subjects belonging to seven experience levels were computed and the data was subjected to the Analysis of Variance to find out (the F-ratio between them) if the differences in the mean scores were significant or not. The results thus obtained are presented in the Table 3.18.

Table 3.18

F-Ratio between the Mean Scientific Attitude Scores of the Subjects Belonging to Various Experience Levels

S.No.	*Experience level*	*Strength (N)*	*Mean*	*S.D.*	*F-ratio*
1.	1-5 years	106	146.01	86.32	
2.	6-10 years	92	190.05	91.50	
3.	11-15 years	73	169.01	86.36	
4.	16-20 years	53	185.38	90.36	F=3.51**
5.	21-25 years	55	197.87	82.31	
6.	26-30 years	33	194.54	83.69	
7.	31-35 years and above	09	166.87	74.55	

** Significant at 0.01 level.

Table 3.18 reveals that while the subjects belonging to the experience level of 21-25 years scored the highest on the Scientific Attitude Scale, the subjects belonging to the experience level of 1-5 years scored the least. The mean Scientific Attitude scores generally increased with levels of experience of the subjects with certain exceptions. The mean scores of the subjects increased with their levels of experience upto 21-25 years barring the experience level of 6-10 years and the mean scores decreased from the experience level of 21-25 years and there on. The F-ratio (3.51) found between the mean scores of the subjects belonging to various experience levels was statistically significant at 0.01 level.

The t-ratios that were found out between the differences in the means of various pairs of experience levels are presented in the Table 3.19.

Table 3.19 reveals that:

1. (a) Subjects belonging to the experience level of 1-5 years scored the least on the Scientific Attitude Scale;

   (b) Subjects belonging to the experience level 1-5 years scored significantly less on the Scientific Attitude Scale than the subjects belonging to the experience levels, 6-10 years, 16-20 years, 21-25 years as their mean differences were found to be significant at 0.01 level;

   (c) the mean differences between the Scientific Attitude scores of the subjects belonging to the experience levels of 1-5 years and 11-15 years and l-5 years and 31-35 years were not found to be statistically significant;

2. (a) Subjects belonging to the experience level of 6-10 years scored higher on the Scientific Attitude Scale than the subjects belonging to the experience levels of 11-15 years, 16-20 years and 31-35 years and above. But the mean differences among them were not found to be statistically significant;

Table 3.19

Differences in the Mean Scientific Attitude Scores of the Subjects with Various Levels of Experience

	6 - 10 yrs.	*11 - 15 yrs.*	*16 - 20 yrs.*	*21 - 25 yrs.*	*26 - 30 yrs*	*31- 35 yrs. and above*
1-5 years	44.04** (6-10 yrs)+	23.00 (11-15 yrs)+	39.37** (16-20 yrs) +	51.86** (21-25 yrs) +	48.53** (26-30 yrs) +	20.86 (31-35 yrs)+
6-10 years		21.04 (6-10 yrs)+	4.67 (6-10 yrs)+	7.82 (21-25 yrs)+	4.49 (26-30 yrs)+	23.18 (6-10 yrs)+
11-15 years			16.37 (16-20yrs)+	28.88 (21-25 yrs)+	25.53 (26-30 yrs)+	2.14 (11-15 yrs)+
16-20 yeas				12.49 (21-25 yrs)+	9.16 (26-30 yrs)+	18.51 (16-20 yrs)+
21-25 years					3.33 (21-25 yrs)+	31.00 (21-25 yrs)+
26-30 years						27.67 (26-30 yrs)+

** Significant at 0.01 level.

* Significant at 0.05 level.

( )+ The group shown in parantheses with plus mark got the higher mean score than the other group under comparison.

(b) Subjects belonging to the experience level of 6-10 years scored less on the Scientific Attitude Scale than the subjects belonging to the experience levels of 21-25 years and 26-30 years. But the mean differences among them were not found to be statistically significant;

3. (a) Subjects belonging to the experience level of 11-15 years scored less on the Scientific Attitude Scale than the subjects belonging to the experience levels 16-20 years, 21-25 years and 26-30 years. But the mean differences among them, however, were not found to be statistically significant:

(b) Subjects belonging to the experience level of 11-15 years scored higher on the Scientific Attitude Scale than the subjects belonging to the experience level of 31-35 years and above. But the mean differences found between them was not significant statistically.

4. (a) Subjects belonging to the experience level of 16-20 years scored less on the Scientific Attitude Scale than the subjects belonging to the experience levels of 21-25 years and 26-30 years. But the mean difference between them were not found to be statistically significant;

(b) Subjects belonging to the experience level of 16-20 years scored higher on the Scientific Attitude Scale than the subjects belonging to the experience level of 31-35 years and above. The mean difference between them was not significant statistically;

5. Subjects belonging to the experience level of 21-25 years scored higher on the Scientific Attitude Scale than the subjects belonging to the experience levels of 26-30 years and 31-35 years and above. The mean differences among them were not found to be statistically significant; and

6. Subjects belonging to the experience level of 26-30 years scored higher on the Scientific Attitude Scale than the

subjects belonging to the experience level of 31-35 years and above. But the mean difference between them was not significant statistically.

The results thus arrived at are as follows:

1. Subjects belonging to the experience level of 1-5 years scored the least on the Scientific Attitude Scale.
2. Subjects belonging to the experience level of 26-30 years scored the highest on the Scientific Attitude Scale.
3. Mean Scientific Attitude scores increased with the levels of experience to certain level and then decreased.

Thus the levels of experience of the subjects influenced their Scientific Attitude scores. And so the null hypothesis of no significant difference between the duration of the subjects' experience and their mean levels of Scientific Attitude gets rejected.

6.B. Discussion

The review of researches presented in the Chapter-II does not shed any light on the dimension highlighting the relationship between the Experience of Academics and Research Scholars on the scores on Scientific Attitude Scale.

The research under report generally yields the positive association between experience and Scientific Attitude scores with an exception of the Scientific Attitude score slightly declining at the fag end of the career of the Academics.

The positive association between the Age and Scientific Attitude score and Income and Scientific Attitude score have already been discerned and discussed. The length of Experience is more akin to Age and Income. In fact Age, Income and Experience are interrelated and could be conceived to have an identical influence on Scientific Attitude scores. The Academics increase their knowledge with

Experience and the Experience underlines the pursuit of Scientific Research for longer durations and both of them would contribute to the positive association between Experience and scores on Scientific Attitude Scale.

The decline of Scientific Attitude scores at the fag end of the careers of the Academics underscores the possibility of the Academics moving out of their Academic pursuits to Administrative positions and in developing disinterest in the Academic pursuits as their superannuation approaches. This gets support even from the learning curve declining after a particular age. Thus we find a general positive association between Experience and Scientific Attitude, the curve declining towards the end of the career of the Academics.

7.A. Hypothesis (7)

There will be no significant Differnce Between the Mean Scientific Attitudes of Academic and Professional Degree Holders.

A null hypothesis of no significant difference between the mean Scientific Attitudes of Academic and Professional degree holders was formulated.

The sample was divided on the basis of Academic and Professional Degrees held by the subjects. The "Means" and "Standard Deviations" of the Scientific Attitude scores of Academic degree holders and professional degree holders were calculated. The t-test was computed to find out if the mean difference was significant or not. The results that were found out are presented in the Table 3.20.

Table 3.20

T-Ratio between the Mean Scientific Attitude Scores of Academic and Professional Degree Holders

Types of the degree	*Strength (N)*	*Mean*	*S.D.*	*t-ratio*
Academic degree holders	311	162.67	89.63	t=0.75
Professional degree holders	198	168.84	92.09	

Table 3.20 reveals that while Professional degree holders scored a mean score of 168.84 on the Scientific Attitude Scale, the Academic degree holders scored a mean score of 162.67. The difference in their mean scores was not found to be statistically significant. And so, it could be said that Academic degree holders and Professional degree holders do not differ in Scientific Attitude.

Thus the null hypothesis of no significant difference between the mean Scientific Attitudes of Academic degree and Professional degree holders is not rejected.

7.B. Discussion

The review of research presented does not yield any evidence in the area of the nature of degree held (Academic and Professional) and the scores on Scientific Attitude Scale.

The present research does not show any significant difference in the mean Scientific Attitude scores on the basis of the nature of degree held.

Generally it has been established that the Academic degree holders are more Liberal while the Professional degree holders are more Conservative (Herbert Shepard, 1952). But since the sample includes more number of oriental languages experts in the sample of Academic degree holders their scores are not found to be significantly higher (though high) than that of the Professional degree holders. "The differences being non significant" could be attributed to sampling procedures. Thus we find in the present Investigation that though the Academic degree holders are found to score less on the Scientific Attitude Scale than the Professional degree holders, their differences are not found to be statistically significant.

8.A Hypothesis (8)

There will be no Significant Relationship in the mean levels of Scientific Attitude of Subjects because of the differences in their Political Ideological Backgrounds.

A null hypothesis of no significant relationship between the political ideological background of the subjects and their mean levels of Scientific Attitude was formulated.

The sample thus classified on the political ideological background of the subjects is as follows:

(i) R.S.S. (N=22)

(ii) B.J.P. (N=33)

(iii) Congress (N=92)

(iv) Janatadal (N=28)

(v) B.S.P. (N=5)

(vi) C.P.I.(N=4)

(vii) C.P.M.(N=14)

(viii) C.P.I.(ML) (N=23)

(ix) Any other party (N=74)

As the number in some political ideologies was meagre, to arrive at dependable conclusions, it was thought advisable to club some of the like minded ideologies into one group. Thus the R.S.S. and B.J.P. were clubbed together and named B.J.P./ R.S.S. The Janatadal and B.S.P. were clubbed together and named B.S.P./Janatadal. The subiects belonging to Political ideologies of C.P.I, and C.P.M. were clubbed together named as C.P.I/C.P.M. group. The subjects belonging to the group of "Any other party" were excluded from the analysis as they had no coherence in their political ideologies. While the other groups could be identified with a single political ideology, "Any other group" could not be identified with a single political ideology; so that group was excluded. After clubbing and eliminating, the sample thus classified for the purpose of the present investigation is as follows:

(i) BJP/RSS (N=55)

(ii) Congress (N=92)

(iii) BSP/Janatadal (N=33)

(iv) CPI/CPM (N=18)

(v) CPI(ML) (N=23)

The means and standard deviations of the Scientific Attitude scores of the subjects belonging to various political ideologies were computed. And the data was subjected to Analysis of Variance to find if the differences in the mean scores were statistically significant or not. The data thus analysed statistically are presented in the Table 3.21.

Table 3.21

F-Ratio between the Mean Scientific Attitude Scores of the Subjects Belonging to Various Political Ideologies.

S. No.	*Type of Political Ideology*	*Strength (N)*	*Mean*	*S.D*	*F-ratio*
1.	BJP/RSS	55	135.78	97.50	
2.	Congress	92	135.42	89.54	
3.	BSP/Janatadal	33	177.64	80.14	F=4.41**
4.	CPI/CPM	18	189.40	76.27	
5.	CPI (ML)	23	203.12	104.11	

** Significant at 0.01 level.

Table 3.21 reveals that while the subjects belonging to CPI(ML) ideological background occupied the first place with a mean score of 203.12; the subjects belonging to the ideology of CPI/CPM obtained the second place with a mean score of 189.40; subjects belonging to the ideology of BSP/Janatadal got the third place with a mean score of 177.64; subjects belonging to the ideology of BJP obtained the fourth place with a mean score of 135.78 and the subjects belonging to the ideology of Congress secured the last place with a mean score of 135.42. The F-value (4.41) obtained between their mean scores was found to be statistically significant at 0.01 level. So it could be said that there were significant differences between their mean scores. Thus t-tests were computed to find out the significant differences between the various pairs of means. The results thus obtained are presented in the Table 3.22.

Table 3.22

Differences between the Mean Scientific Attitude Scores of the Subjects Belonging to Various Political Ideologies

	Congress	*BSP/Janatadal*	*CPI/CPM*	*CPI(ML)*
BJP/RSS	0.36 (BJP/RSS)+	41.86* BSP/Janatadal)+	53.62* (CPI/CPM)+	67.34** (CPI(ML))+
Congress		42.22* (BSP/Janatadal)+	53.98* (CPI/CPM)+	67.70** CPI (ML))+
BSP/Janatadal			11.76 (CPI/CPM)+	25.48 (CPI(ML))+
CPI/CPM				13.72 (CPI(ML))+

** Significant at 0.01 level.

* Significant at 0.05 level.

( )+ The group shown in the parantheses got the higher mean score than the other group under comparison.

Table 3.22 reveals the following:

1. (a) Subjects belonging to the ideology of BJP/RSS scored higher on the Scientific Attitude Scale than the subjects

belonging to the ideology of Congress. But the mean difference was not found to be statistically significant.

(b) Subjects belonging to the ideology of BJP/RSS scored significantly less on the Scientific Attitude Scale than the subjects belonging to the ideologies of BSP/ Janatadal, CPI/CPM and CPI(ML) groups as the mean differences were found to be statistically significant.

2. Subjects belonging to the ideology of Congress scored significantly less on the Scientific Attitude Scale than the subjects belonging to the ideologies of BSP/ Janatadal, CPI/CPM and CPI(ML) groups as their mean differences were found to be statistically significant.

3. Subjects belonging to the ideology of BSP/Janatadal scored less on the Scientific Attitude Scale than the subjects belonging to the ideologies of CPI/CPM and CPI(ML) groups. But the differences in their mean scores were not found to be statistically significant.

4. Subjects belonging to the political ideology of CPI/CPM group scored less on the Scientific Attitude Scale than the subjects belonging to ideology of CPI(ML) group. But these mean difference was not found to be statistically significant.

Thus, it could be said that the political ideologies of the subjects had an association with their Scientific Attitude scores. So the null hypothesis of no significant relationship in the mean levels of Scientific Attitude of subjects because of the differences in their political ideological backgrounds is rejected.

But the present investigator was equally interested in finding out if there could be any differences on each dimension of the Scientific Attitude Scale. And so the data was subjected to Analysis of Variance, dimensions-wise, and F-ratios were computed.

Whenever the F-ratios were significant, the mean scores among the pairs of groups were tested through t-tests. And the results thus arrived at are presented in the following paragraphs through tables under separate captions. The mean differences, whenever they were significant, have been reported in this thesis.

8.A.1. Rationality

The results pertaining to the Rationality dimension are presented in the Table 3.22.1.

Table 3.22.1

F-Ratio between the Mean Scientific Attitude Scores of the Subjects on the Dimension of Rationality in Relation to their Political Ideological Background

Type of political ideology	*Strength (N)*	*Mean*	*S.D.*	*F-ratio*
BJP/RSS	55	28.90	25.94	
Congress	92	38.06	28.43	
BSP/Janatadal	33	49.65	30.36	F=3.96**
CPI/CPM	18	60.59	32.11	
CPI/(ML)	23	54.53	34.88	

** Significant at 0.01 level.

Table 3.22.1 reveals that the subjects belonging to the ideology of CPI/CPM got the highest score on the dimension of Rationality with a mean score of 60.59 while the subjects belonging to the ideology of BJP/RSS got the least score with a mean score of 28.90. The mean scores of the subjects belonging to other political ideologies varied in between the Range 28.90-60.59. The F-ratio (3.96) obtained between their mean scores was found to be statistically significant at 0.01 level. Thus it could be said that the political ideology of the subjects had an association with the Scientific Attitude score on the dimension of Rationality.

So the differences that were found between the mean scores were tested through t-tests and the results obtained are presented in the Table 3.22.2.

Table 3.22.2

Differences in the Mean Scientific Attitude Scores of the Subjects on the Dimension of Rationality in Relation to their Political Ideological Background

	Congress	*BSP/Janatadal*	*CPI/CPM*	*CPI(ML)*
BJP/RSS	9.76 (Congress)+	21.35** (BSP/Janatadal)+	32.29** (CPI/CPM)+	26.23** (CPI(ML))+
Congress		11.59 (BSP/Janatadal)+	22.53** (CPI/CPM)+	16.47** (CPI(ML))+
BSP/Janatadal			10.94 (CPI/CPM)+	4.88 (CPI(ML))+
CPI/CPM				6.06 (CPI/CPM)+

** Significant at 0.01 level.
( )+ The group shown in parantheses with plus marks got the higher mean score than the other group under comparison.

Table 3.22.2 reveals that:

1. The subjects belonging to the political ideologies of BSP/ Janatadal, CPI/CPM and CPI(ML) groups scored significantly higher on the dimension of Rationality than the subjects belonging to BJP/RSS group as their mean differences were found to be statistically significant at 0.01 levels.

2. Subjects belonging to the political ideologies of CPI/ CPM and CPI(ML) groups scored significantly higher on the dimension of Rationality than the subjects belonging to Congress. The mean differences among them were found to be statistically significant at 0.01 levels.

8.A.2 Curiosity

The results pertaining to the dimension of Curiosity are presented in the Table 3.22.3.

Table 3.22.3

F-Ratio between the Mean Scientific Attitude Scores of the Subjects on the Dimension of Curiosity in Relation to their Political Ideological Background

Type of Political Ideology	*Strength (N)*	*Mean*	*S.D.*	*F-ratio*
BJP/RSS	55	37.02	29.57	
Congress	92	30.32	25.02	
BSP/Janatadal	33	36.90	31.81	F=1.67
CPI/CMP	18	43.97	24.06	
CPI/(ML)	23	40.45	28.06	

Table 3.22.3 reveals that, while the subjects belonging to CPI/CPM group scored the highest on the dimension of Curiosity with a mean score of 43.97, the subjects belonging to Congress scored the least with a mean score of 30.32. The mean scores of other political ideologies varied in between the range of 30.32 - 43.97. The F-ratio (1.67) computed between their mean scores was not found to be statistically significant.

Thus it gets established that the political ideological background of the sample does not have any association with their mean Scientific Attitudes scores on the dimension of Curiosity. Thus the subjects belonging to various political ideologies do not differ in their mean scores on Curiosity dimension.

8.A.3. Openmindedness

The results obtained with regard to the dimension of Openmindedness are presented in the Table 3.22.4.

Table 3.22.4

F-Ratio between the Mean Scientific Attitude Scores of the Subjects on the Dimension of Openmindedness in Relation to their Political Ideological Background

Type of Political Ideology	*Strength (N)*	*Mean*	*S.D.*	*F-ratio*
BJP/RSS	55	22.55	28.72	
Congress	92	18.40	24.27	
BSP/Janatadal	33	20.96	24.72	F=1.30
CPI/CPM	18	17.31	25.16	
CPI (ML)	23	32.27	24.49	

Table 3.22.4 reveals that the subjects belonging to the political ideology of CPI(ML) got the highest score on the dimension of Openmindedness with a mean score of 32.27. The subjects belonging to the political ideology of CPI/CPM got the least score on the dimension of Openmindedness with a mean score of 17.31. The mean scores of other groups fell in between 17.31 and 32.27. The F-ratio (1.30) obtained between their mean scores was not found to be statistically significant.

Thus, it gets established that the political ideology of the subjects does not have any association with their mean Scientific Attitude scores on the dimension of Openmindedness.

8.A.4. Aversion to Superstitious Beliefs

The results pertaining to the dimension of Aversion to Superstitious Beliefs are presented in the Table 3.22.5 and 3.22.6.

Table 3.22.5

F-Ratio between the Mean Scientific Attitude Scores of the Subjects on the Dimension of Aversion to Superstitious Beliefs in Relation to their Political Ideological Background

Type of Political Ideology	*Strength (N)*	*Mean*	*S.D.*	*F-ratio*
BJP/RSS	55	6.14	17.82	
Congress	92	11.10	13.97	
BSP/Janatadal	33	19.40	8.39	F=6.61**
CPI/CPM	18	22.48	8.93	
CPI (ML)	23	25.05	8.58	

** Significant at 0.01 level.

Table 3.22.5 reveals that while the subjects belonging to political ideology of CPI(ML) scored the highest on the dimension of Aversion to Superstitious Beliefs with a mean score of 25.05, the subjects belonging to the BJP/RSS political ideology scored the least with a mean score of 6.14. The mean scores of other groups varied in between the range of 6.14-25.05. And F-ratio found in between their mean scores was found to be statistically significant at 0.01 level. So it can be said that the political ideology of the subjects has an association with the mean Scientific Attitude scores on the dimension of Aversion to Superstitious Beliefs.

The differences that were observed in the mean scores were tested through t-test and the results thus obtained are presented in the Table 3.22.6.

Table 3.22.6 reveals that:

1. The subjects belonging to the political ideologies of Congress, BSP/Janatadal, CPI/CPM and CPI(ML) scored significantly higher on the dimension of Aversion to Superstitious Beliefs than the subjects belonging to the

Table 3.22.6

Differences in the Mean Scientific Attitude Scores of the Subjects on the Dimension of Aversion to Superstitious Beliefs in Relation to their Political Ideological Background.

	Congress	*BSP/Janatadal*	*CPI/CPM*	*CPI(ML)*
BJP/RSS	4.96*	13.26**	16.34**	18.91**
	(Congress)+	(BSP/Janatadal)+	(CPI/CPM)+	(CPI(ML))+
Congress		8.30**	11.38**	13.95**
		(BSP/Janatadal)+	(CPI/CPM)+	(CPI(ML))+
BSP/Janatadal			3.08	5.65
			(CPI/CPM)+	(CPI(ML)) +
CPI/CPM				2.57
				(CPI (ML)) +

** Significant at 0.01 level.
* Significant at 0.05 level.
( )+ The group shown in parantheses with a plus mark got the higher mean score than the other group under comparison.

ideology of BJP/RSS as their mean differences were found to be statistically significant.

2. Subjects belonging to the political ideologies of BSP/ Janatadal. CPI/CPM and CPI(ML) scored significantly higher on the dimension of Aversion to Superstitious Beliefs than the subjects belonging to the political ideology of Congress as their mean differences were found to be statistically significant.

8.A.5. Objectivity and Intellectual Honesty

The results pertaining to the dimension of Objectivity and Intellectual Honesty are presented in the Table 3.22.7.

Table 3.22.7 reveals that while the subjects belonging to the political ideology of CPI(ML) scored the highest on the dimension of Objectivity and Intellectual Honesty with a mean

Table 3.22.7

F-Ratio between the Mean Scientific Attitude Scores on the Dimension of Objectivity and Intellectual Honesty in Relation to their Political Ideological Background

Type of Political Ideology	*Strength (N)*	*Mean*	*S.D.*	*F-ratio*
BJP/RSS	55	22.07	20.32	
Congress	92	20.78	24.29	
BSP/Janatadal	33	26.49	18.00	F=1.05
CPI/CPM	18	17.93	21.14	
CPI (ML)	23	27.26	25.57	

score of 27.26, the subjects belonging to the political ideology of CPI/CPM scored the least with a mean score of 17.93. The mean scores of other groups fell in between 17.93 and 27.26. The F-ratio (1.05) obtained between their mean scores was not found to be statistically significant. Thus it is established that the political ideology of the subjects does not have any association with their mean Scientific Attitude scores on the dimension of Objectivity and Intellectual Honesty.

8.A.6. Suspended Judgement

The results pertaining to the dimension of Suspended Judgement are presented in the Table 3.22.8.

Table 3.22.8 reveals that the subjects belonging to the political ideology of CPI/CPM scored the highest on the dimension of Suspended Judgement with a mean score of 24.70 and the subjects belonging to the political ideology of BJP/RSS scored the least with a mean score of 16.06. The mean scores of other groups ranged between 16.06-24.70. The F-ratio (1.83) found between these mean scores was not statistically significant. Thus it gets established that the political ideology of the subject does not have any significant effect on their mean Scientific Attitude scores on the dimension of Suspended Judgement.

Table 3.22.8

F-Ratio between the Mean Scientific Attitude Scores of the Subjects on the Dimension of Suspended Judgement in Relation to their Political Ideological Background

Type of Political Ideology	*Strength (N)*	*Mean*	*S.D.*	*F-ratio*
BJP/RSS	55	16.06	13.08	
Congress	92	18.11	15.19	
BSP/Janatadal (op)	33	23.14	12.56	F=1.83
CPI/CPU	18	24.70	11.71	
CPI (ML)	23	20.27	16.86	

8.A.7. Theory Building

The results pertaining to the dimension of Theory Building are presented in the Table 3.22.9.

Table 3.22.9

F-Ratio between the Mean Scientific Attitude Scores of the Subjects on the Dimension of Theory Building in Relation to their Political Ideological Background

Type of Political Ideology	*Strength (N)*	*Mean*	*S.D.*	*F-ratio*
BJP/RSS	55	4.56	6.47	
Congress	92	2.42	6.97	
BSP/Janatadal (op)	33	2.51	8.19	F = 1 .12
CPI/CPM	18	5.06	9.03	
CPI (ML)	23	3.28	6.88	

Table 3.22.9. reveals that while the subjects belonging to the political ideology of CPI/CPM scored the highest on the dimension of Theory Building with a mean score of 5.06, subjects belonging to the political ideology of Congress scored the least with a mean score of 2.42. The means of other subjects fell in between the range of 2.42-5.06. The F-ratio (1.12) obtained between these mean scores was not found to be statistically significant. Thus it gets established that the political ideology of the subjects does not have any association with their mean Scientific Attitude scores on the dimension of Theory Building.

The results presented under the captions 8.A., 8.A.1., A.2., 8.A.3., 8.A.4., 8.A.5., 8.A.6. and 8.A.7 thus reveal that:

1. Political ideology of the subjects has a significant association with their mean Scientific Attitude scores on the dimension of Rationality and Aversion to Superstitious Beliefs.
2. Political ideology of the subjects does not effect the mean Scientific Attitude scores on the dimensions of Curiosity, Openmindedness, Objectivity and Intellectual Honesty, Suspended Judgement and Theory Building.
3. Political ideology of the subjects has a significant association with the mean scores on the Scientific Attitude Scale.

8.B. Discussion

The researches reviewed in Chapter-II do not give any information covering the relationship between the Political ideological background of the Academics and their scores on Scientific Attitude Scale. But the results of the research under report reveal a significant difference in the mean Scientific Attitude scores in relation to the political bent of the Academics. It has been established that while the Leftists with the ML ideology at the top have scored the highest, the Congress and the BJP/RSS sympathisers scored the least. The BSP/Janatadal were found much closer to the Leftists so much so that there was no significant difference between their scores on Scientific Attitude Scale. Significant differences were found between (i) the Congress and BSP/Janatadal, (ii) Congress and CPI/CPM and CPI(ML) groups, (iii) BJP/RSS and BSP/Janatadal and (iv) BJP/RSS and CPI/CPM and CPI(ML) groups. There were no significant differences between BSP/Janatadal and CPI/CPM and CPI(ML) groups and in between the communist groups.

It is understandable that BJP/RSS group which swears by the hoary past of the Indian Culture seething in relative religious orthodoxy, scores less on the Scientific Attitude Scale. It is not surprising that the votaries of BJP/RSS have scored the least in the dimensions of Rationality and Aversion to Superstitious Beliefs. But it is surprising that the Academics and Research Scholars who profess allegiance to Congress party which is supposed to stand for secular and rational values, should score least on the Scientific Attitude Scale. This obviously is a drift of the Congress Party from its professed ideals.

Academics and Research Scholars who claim closeness to the Left parties have scored the highest on the Scientific Attitude Scale.

9.A. Hypothesis (9)

There will be no significant difference in the Scientific Attitudes of Subjects because of their levels of Political Activism.

A null hypothesis of no significant difference between Scientific Attitudes of the subjects because of the differences in their levels of Political activism was formulated.

The sample thus obtained for this purpose was classified on the basis of three levels of political activism namely Cadre/ worker level, Organisation level and Theoretician level and then the data was subjected to Analysis of Variance. The results thus found are presented in the Table 3.23.

Table 3.23

F-Ratio between the Mean Scientific Attitude Scores of the Subjects belonging to Various Levels of Political Activism

Level of Political Activism	*Strength (N)*	*Mean*	*S.D.*	*F-ratio*
Cadre/Worker level	28	110.91	88.37	
Organisational level	21	116.64	99.32	F.7 .92**
Theoretician level	104	175.02	87.79	

** Significant at 0.01 level.

Table 3.23 reveals that while the subjects belonging to the Theoretician level of the Political activism secured the first place on the Scientific Attitude Scale with a mean score of 175.02, subjects belonging to the Organisational level got the second place with a mean score of 116.64 and the subjects belonging to the Cadre level obtained the third place with a mean score of 110.91. The mean scores fell in line with the hierarchy of the level of Political activism. The F-ratio (7.92) obtained between their mean Scientific Attitude scores was found to be statistically significant at 0.01 level.

Thus it gets established that the level of Political activism of the subjects has also association with their mean Scientific Attitude scores.

The mean differences that were observed between the pairs of means were tested through t-tests and the results thus obtained are presented in the Table 3.24.

Table 3.24

Differences in the Mean Scientific Attitude Scores of the Subjects belonging to Various Levels of Political Activism

	Organisation level	*Theoretician level*
Cadre/worker level	5.73 (organisation level)+	64.11** (Theoretician level)+
Organisation level		58.38** (Theoretician level) +

** Significant at 0.01 level.
( ) + The group shown in parantheses with plus marks got the higher mean score than the other group under comparison.

Table 3.24 reveals that:

1. (a) Subjects belonging to the Cadre level scored less on the Scientific Attitude Scale than the subjects belonging to the Organisation level, and their mean difference was not found to be statistically significant.

   (b) Subjects belonging to the Cadre level significantly scored less on the Scientific Attitude Scale than the subjects belonging to the Theoretician level as their mean difference was found to be statistically significant at 0.01 level.

2. Subjects belonging to the Organisational level significantly scored less on the Scientific Attitude Scale than the subjects belonging to the Theoretician level as their mean difference was found to be statistically significant at 0.01 level.

The results presented in the Tables 3.23 and 3.24 reveal that:

1. the level of Political activism of the subjects has an influence on their Scientific Attitude scores; and
2. the subjects belonging to the Theoretician and Organisation levels significantly scored higher on the

Scientific Attitude Scale than the subjects belonging to the Cadre level.

Thus the null hypothesis of no significant difference in the mean Scientific Attitudes of subjects because of their levels of Political activism gets rejected.

9.B. Discussion

Generally the people professing allegiance to any particular ideology operate at various levels. In the research under report, three levels were conceived in a hierarchy. They were worker level, organiser level and theoretician level. A theoretician normally theorises whatever he professes on a rational basis. Such theorisation, it was conceived, would make him more rational within the framework of his theory and so would score more on a Scientific Attitude Scale.

The second level is that of an Organiser. He devotes more time in the organisation aspect rather than on theorisation.

The worker operates at the third level. Even though he may be an Academic in a University he executes whatever he is asked to.

It was believed that the operation at these three levels would display differences in the Scientific Attitude scores. The result of the present research have supported such a hypothesis and the scores obtained on the Scientific Attitude Scale at these three levels fell in the same hierarchy, the theoretician scoring the highest, and the worker scoring the least. The differences between the scores of the theoretician and the organiser and the scores between the theoretician and the worker were found to be significant. However, the differences in the scores of the organiser and the worker were not significant. This could be explained by the fact that the worker too is working in college or a University along with the organiser. Both of them are almost equally qualified academically and so the differences were not found to be significant.

The fact that the theoretician has scored high might as well be due to his seniority coupled with his ability to theorise whatever he professes.

Thus it is concluded that the Academics who theorise score more on the Scientific Attitude Scale than the Academics who organise and execute.

10. A. Hypothesis (10)

There will be no significant difference between the mean scientific attitudes of the subjects who had backgrond of some political philosophy and those who had no background of any political philosophy.

A null hypothesis of no significant difference between the mean Scientific Attitudes of the subjects who had background of some political philosophy and those who had no background of any political philosophy, was formulated.

The sample was classified on the basis of the subjects who had background of some political philosophy and who had no background of political philosophy. The means and standards deviations of the Scientific Attitude scores of the said groups were computed and the mean difference between the two groups was tested through t-test. The results thus obtained are presented in the Table 3.25.

Table 3.25 reveals that the subjects who had no background of any political philosophy scored higher on the Scientific Attitude Scale than the subjects who had the background of some political philosophy. The t-value (1.34) obtained between their mean Scientific Attitude scores however was not found to be statistically significant.

Table 3.25

T-Ratio between the Mean Scientific Attitude Scores of the Subjects who had Background of Some Political Philosophy and who had no Background of any Political Philosophy

S. No.	*Type of Subjects*	*Strength (N)*	*Mean*	*S.D.*	*t-ratio*
1.	Subjects who had background of some political philosophy	295	1.60.53	91.98	t=1.34
2.	Subjects who had no background of any Political Philosophy	214	171.33	88..38	

Thus the null hypothesis of no significant difference between the Scientific Attitudes of the subjects who had background of some political philosophy and those who had no background of any political philosophy is not rejected.

10.B. Discussion

Generally it is conceived that Academics who are sensitive to Socio-economic issues of the society take to political ideologies and their ideological commitment is claimed to be a reflection of their sensitivity. It is claimed that any sensitive person will be attracted to political ideology and that the ideology would foster a sort of rationality of its own.

The votaries of the ideology do not have appreciative stance over politically non committed Academics and Research Scholars.

Does the commitment to some Political ideology imply a greater score on the Scientific Attitude Scale than the non-commitment?

The research under report sought to find out an answer to the poser.

The result reveal that ideologically uncommitted Academics and Research Scholars scored higher on the Scientific Attitude Scale than the ideologically committed Academics and Research Scholars though the differences were not found to be statistically significant. It means that the Academics and Research Scholars committed to some or other Political ideology and the Academics and Research Scholars who are not committed to any Political ideology do not differ in their Scientific Attitude.

Results and Conclusions

The present investigation found the following results:

1. Results

1. By and large there was no significant difference in the Scientific Attitudes of Males and Females on all the dimensions of Scientific Attitude except in the dimension of Theory Building.

2. Academics scored significantly higher on all the dimensions of Scientific Attitude Scale except the dimension of Aversion to Superstitious Beliefs. Academics and Research Scholars did not differ on the dimension of Aversion to Superstitious Beliefs.

3. Age levels of the subjects have a positive association with their Scientific Attitude. The higher the age level, the higher was the Scientific Attitude.

4. Income levels of the subjects have a positive association with their Scientific Attitude.

5. (a) The content background of the subjects influenced the Scientific Attitude of the subjects on the dimensions of "Curiosity", "Open-mindedness", "Aversion to Superstitious Beliefs", "Objectivity-Intellectual Honesty" and on the total dimensions of the Scientific Attitude.

(b) The content background of the subjects does not influence the Scientific Attitude of the subjects on the dimensions of "Rationality". "Suspended Judgement" and "Theory Building".

(c) Subjects belonging to Physical Sciences scored the highest on the Scientific Attitude Scale.

(d) Subjects belonging to Physical Sciences scored significantly higher on the Scientific Attitude Scale than the subjects belonging to Humanities, Biological Sciences and Health Sciences.

(e) Subjects belonging to Social Sciences scored significantly higher on the Scientific Attitude Scale than the subjects belonging to Biological Sciences.

(f) Subjects belonging to Engineering and Mathematics groups scored significantly higher on the Scientific Attitude Scale than the subjects belonging to Biological Sciences.

(g) Subjects belonging to Biological Sciences scored the least on the Scientific Attitude Scale.

6. By and large mean Scientific Attitude of the subjects increased with their levels of experience to certain level and then decreased.

7. Academic degree holders and Professional degree holders do not differ in their Scientific Attitude.

8. (a) Political ideology of the subjects has a significant association with their mean Scientific Attitude on the dimensions of "Rationality", "Aversion to Superstitious Beliefs" and on the total dimensions of the Scientific Attitude.

(b) Political ideology of the subjects does not effect the mean Scientific Attitude score on the dimensions of "Curiosity", "Open-mindedness", "Objectivity-

Intellectual Honesty", "Suspended Judgement", and "Theory Building".

(c) Subjects belonging to the ideology of BJP/RSS scored significantly less on the Scientific Attitude Scale than the subjects belonging to the ideologies of BSP/ Janatadal, CPI/CPM, and CPI(ML) groups.

(d) Subjects belonging to the ideology of Congress scored significantly less on the Scientific Attitude Scale than the subjects belonging to the ideologies of BSP/ Janatadal, CPI/CPM and CPI(ML) groups.

9. (a) The level of Political activism of the subjects has an influence on their Scientific Attitude; and

(b) the subjects belonging to Theoretician and Organisational levels significantly scored higher on the Scientific Attitude Scale than the subjects belonging to the Cadre level.

10. Subjects committed to some or other Political ideology and the subjects who are not committed to any other Political ideology do not differ significantly in their Scientific Attitude.

The present investigation has arrived at the following conclusions.

2. Conclusions

1. Males and Females do not differ in their Scientific Attitude.

2. Academics and Research Scholars differ in their Scientific Attitude.

3. Scientific Attitudes of the subjects increased with their age levels.

4. Scientific Attitude of the subjects increased with their income levels.

5. The content background of the subjects influenced their Scientific Attitude. The Physicists are found at the top on the scores on Scientific Attitude Scale followed by Mathematicians, Social Scientists, Engineers, Health Scientists and Academics working in the faculties of Arts and Humanities and finally the Botanists and Zoologists.
6. Scientific Attitude of the subjects increased with their level of experience.
7. Academic degree holders and Professional degree holders do not differ in their Scientific Attitude.
8. Political ideologies of the subjects has an association with their Scientific Attitude. Subjects who claim closeness to the left parties have scored the highest on the Scientific Attitude Scale and the subjects who claim to be rightists scored the least on the Scientific Attitude Scale.
9. The level of Political activism of the subjects too has an influence on their Scientific Attitude. Subjects who claim to be as Theoreticians scored more on the Scientific Attitude Scale than the subjects belonging to the Organisation and Cadre levels.

Subjects committed to some or other Political ideology and the subjects who are not committed to any other Political ideology do not differ in their Scientific Attitude.

Suggestions for Further Studies

The study under report has generated the following problems, which deserve research attention. They are:

1. The Scientific Attitude Scales are to be constructed and standardised for the students at secondary and tertiary levels.
2. It has been observed earlier that there is an association between the content area of the subjects and the scores

on the Scientific Attitude Scale. Candidates from Physical Sciences, Mathematics and Social Sciences have been scoring high and the candidates from certain other subjects very low.

There is a possibility that the width of the content area or lack of it because of concentration in specialised functioning will have positive or not very positive effect on Scientific Attitude scores. Subjects like Philosophy and Philosophy of Science etc., which cover the whole range of thinking may have a positive association with the scores on Scientific Attitude Scale. Whereas specialisations like ophthalmology, skin diseases etc. may not be very conducive to higher scores on Scientific Attitude Scale. Such a hunch needs to be investigated.

Appendices

Appendix 1: BIO-DATA

1. Name

2. Sex

Male		Female	

3. Occupation

4. Place of Work

5. Age (In Completed Years)

6. Religion

Hindu	Muslim	Christian	Sikh	Parsi

7. Caste

8. Monthly Income From all Sources

Less Than 500	Rs.500 - Rs. 1000	Rs. 1000 - 2500

Rs. 2500-5000	Rs. 5000-8000	Rs. 8000 & Above

9. Qualification (Highest)

(i) Academic

Under Graduate	B.A.	B.Com.	B.Sc.

M.A.	M.Com.	M.Sc.	Ph.D

(ii) Professional

M.S.	M.D.	M.Tech.	LL M.	M.Ed	Ph.D.

10. Main Content Background

Humanities & Arts	Social Sciences	Physical Sciences

Biological Sciences	Health Sciences	Engineering	Mathematics

11. Professional Experience (in Completed Years)

5 years & less	10 years	15 years	20 years

30 years	35 years and above

12. My Political Philosophy is Close to

R.S.S. (Shivasena)	B.J.P.	Congress	Janatadal

B.S.P.	CPI	CPM	CPI (ML)	Any other

13. I have been a Political Activist at the Level

Cadre/Worker Level	Leadership at Organisational Level	Theoretician Level

Scientific Attitude Scale

	Agree	Disagree
1. I accept only those things which stand the test of reasoning.		
2. When I am convinced that the approach I have adopted is perfect why would I adopt multiple alternative approaches to people, things and ideas.		
3. As the Truth is unchanging the ideas about the Truth should remain unchanged.		
4. I shall not accept any traditional belief unless I test it.		
5. Obviously I do not feel very comfortable when I listen to, observe, or read evidence contradicting my strong personal opinions.		
6. Any natural calamity is the result of certain natural causes and so to understand the nature of calamity, its antecedents need to be studied.		
7. Critical faculties are instrumental in expanding the horizons of knowledge.		
8. Even the so called superstitions have deep roots in reality which we often fail to take into cognisance.		
9. No authority is infallible. So one should not hesitate to challenge it.		
10. When the findings or theories of Science conflict with religious belief,		

	Agree	Disagree
it is better to accept the religious belief.		
11. It is the mediocre who insists on observation & recording of facts as the wise can formulate confusions even without the crude data.		
12. Humility does not signify confession of ignorance. Even when one is sure that he knows more than others, he need not give up humility.		
13. When the observed phenomenon is self evident, one need not insist for further explanation.		
14. As the old order changes yielding place to new, one needs to be always ready to understand the new.		
15. I find it difficult to resist my curiosity to find out the whys and hows of observed phenomena.		
16. Old evidences related to a problem are more authentic than the most recent ones.		
17. If I were a teacher I would encourage my students to question me as closely as they can.		
18. Quantification which is being employed these days is an untrustworthy mechanistic device.		
19. It is obvious that often we draw final inference first and then collect supporting evidence.		

	Agree	Disagree
20. The scientists unending quest is a part of his desire to understand the reality in its totality.		
21. One need not insist for supporting proof when great people themselves make statement of facts.		
22. As I grow I often find myself revising my opinions and conclusions.		
23. I feel excited when I see a new thing or hear a new idea.		
24. When correct theorization is made, the probabilities get reduced to one.		
25. I do not feel uncomfortable in revising the approaches I formulate towards people, things and ideas.		
26. There are many things in the world which transcend human rationality and so must be accepted.		
27. The traditional views have survived over the centuries because of their validity and so they must be accepted.		
28. It is likely that much of the scientific information we have today will be demonstrated to be inaccurate or inadequate in the future.		
29. Gandhiji has asserted that the Bihar earth quake was a punishment inflicted by the nature on Biharies for the sin of untouchability.		

	Agree	Disagree
30. I do not hesitate to receive any evidence, written or spoken, which opposes my strongest personal opinion.		
31. I hate all substitutions which masquerade as traditions in our country.		
32. Being always critical is not a virtue.		
33. I prefer to accept facts which are rationally explained.		
34. Some, because of their expertise and mastery over a field of human activity are considered to be the authorities. It would be unwise to challenge them.		
35. It is hypocracy to demonstrate humility when one is sure that he knows more than others.		
36. A good scientist always demonstrates concern for observation and recording of facts.		
37. All the possible patterns are already conceived by the great theoreticians I adore. Nothing new which has not already been theorised can emerge which merits renewed understanding.		
38. Seeking further information about a novel situation is considered as a sound approach.		

	Agree	Disagree
39. When a gamut of authoritative evidence relating to a problem is furnished, I prefer to select the most recent one.		
40. We must be concerned with the phenomenon we confront with. It is sheer adventurism to seek to unearth the antecedents of each observed phenomenon.		
41. No meaningful conclusions can be drawn unless the evidence is qualified.		
42. When candid and lucid narrations can make the students properly understand the things to be taught, it is (naive) pedantic to resort to questioning method.		
43. Man has ready the command over the totality of knowledge. Still greater efforts in the directions of knowing more are uncalled for.		
44. I do not feel like drawing final inferences before convincing evidence is collected.		
45. As I am not in the habit of entertaining wrong opinions and defective conclusions, I need not change them.		
46. One need not accept the statements emanating from great men as facts unless they are supported by convincing proof.		

	Agree	Disagree
47. There are always alternative competing probabilities which deserve to be accepted.		
48. I do not feel comfortable when I confront things and ideas which I did not come across earlier.		

APPENDIX 2

LIST OF THE JUDGES

1. Prof. V. Sivaram Prasad, Head, Dept. of Mathematics, Osmania University, Hyderabad.
2. Prof. J. Rama Kantha Rao, Dept. of Mathematics, Osmania University, Hyderabad.
3. Prof. Abhayanker, Dept. of Astronomy, Osmania University.
4. Prof. Thayamma Chiranjeevi, Dept. of Physics, Osmania University, Hyderabad.
5. Dr. Ravi Chandra, Reader, Dept. of Psychology, Osmania University, Hyderabad.
6. Dr. Wilson, Reader, Dept. of Philosophy, Osmania University.
7. Dr. D. Satyanarayana, Lecturer in History, Women's College, Koti, Hyderabad.
8. Dr. Kodanda Ram Reddy, Lecturer in Political Science, Women's College, Koti, Hyderabad.
9. Prof. Rama Melkote, Dept. of Political Science, Women's College, Koti, Hyderabad.
10. Prof. Sivarama Shastri (Rtd) Dept. of Physics, O.U.
11. Dr. Haribabu, Reader, Dept. of Physics, Osmania University.
12. Dr. L. Srinivasa Rao, Reader, Dept. of Physics, O.U.
13. Dr. V. Sudhakar, Lecturer, Cell of Education, O.U.
14. Prof. Upadhyaya, Principal, Nizam College, Hyderabad.
15. Prof. Raghavendar Rao, Dept. of Sociology, O.U.
16. Prof. Thakur, Dept. of Zoology, Osmania University.
17. Prof. C.K. Kokate, Faculty of Pharmaceutical Sciences, Kakatiya University, Warangal.

18. Dr. Rama Kantha Rao, Reader, Dept. of Physics, K.U.
19. Dr. Y.W.S. Prasad, Reader, A.L. College of Education, Guntur.
20. Miss Gita Nagaraj, Lecturer in English, RIE, Banglore.
21. Rajendra Prasad, Engineer, Hyderabad.
22. Dr. P. Samma Reddy, Lecturer, College of Education, Warangal.
23. Dr. P. Sandeep, Reader, College of Education, O.U.
24. Prof. E.V. Sundaram, Dept. of Chemistry, Kakatiya University.
25. Prof. Deshmukh, Dept. of Physics, Kakatiya University.
26. Dr. M.C. Prabhakar, Reader, Dept. of Pharmacy, K.U.
27. Dr. A. Rama Chandraiah, Lecturer in Chemistry, R.E.C. Warangal.
28. Dr. S.V.S.R. Reddy, Lecturer in Physics, R.E.C. Warangal.
29. Dr. I. Ajit Kumar Reddy, Lecturer in Chemistry, R.E.C. Warangal.
30. Prof. Sri Hari Rao, Dept. of ECE, R.E.C. Warangal.
31. Dr. Ramaiah, Asst. Professor, Dept. of Chemistry, R.E.C. Warangal.
32. Dr. A.R.C. Reddy, Lecturer in Physics, R.E.C. Warangal.
33. Dr. T. Bhaskar Rao, Reader in Physics, Kakatiya University, Warangal.

Appendix 3

Data Pertaining to Each Statement of Scientific Attitude Scale

Cumulative Frequency Graph for Item No. 1

I accept only those things which stand the test of reasoning

Pile No.	*Frequency*	*Percentage*	*Cumulative Percentage*
1.	0	0	0
2.	0	0	0
3.	0	0	0
4.	0	0	0
5.	0	0	0
6.	2	6.06	6.06
7.	1	3.03	9.09
8.	5	15.15	24.24
9.	2	6.06	30.30
10.	7	21.21	51.51
11.	16	48.49	100.00

Median = 9.5, Q=1.2. Item Selected.

Cumulative Frequency Graph for Item No. 2

Properly conceived Theoretical Generalizations need not be changed as the New Generalizations Emerge

Pile No.	*Frequency*	*Percentage*	*Cumulative Percentage*
1.	8	25.00	25.00
2.	5	15.63	40.63
3.	4	12.50	53.13
4.	2	6.25	59.38
5.	0	0	59.38
6.	2	6.25	65.63
7.	1	3.12	68.75
8.	2	6.25	75.00
9.	4	12.50	87.50
10.	3	9.38	96.88
11.	1	3.12	100.00

Median = 2.20, Q=3.50. Item Rejected.

Cumulative Frequency Graph for Item No. 3

I shall not Accept any Traditional Belief unless I test it.

Pile No.	*Frequency*	*Percentage*	*Cumulative Percentage*
1.	0	0	0
2.	0	0	0
3.	3	9.09	9.09
4.	1	3.03	12.12
5.	0	0	12.12
6.	4	12.12	24.24
7.	5	15.15	39.39
8.	2	6.06	45.45
9.	3	9.09	54.54
10.	6	18.18	72.72
11.	9	27.28	100.00

Median = 8.00, Q = 2.00. Item Selected.

Cumulative Frequency Graph for Item No. 4

When correct Theorization is made the Probabilities get reduced to one

Pile No.	*Frequency*	*Percentage*	*Cumulative Percentage*
1.	0	0	0
2.	0	0	0
3.	3	10.35	10.35
4.	1	3.45	13.80
5.	2	6.90	20.70
6.	10	34.48	55.18
7.	1	3.45	58.63
8.	2	6.89	65.52
9.	1	3.45	68.97
10.	5	17.24	86.21
11.	4	13.79	100.00

Median = 5.4, Q = 2.05. Item Selected.

Cumulative Frequency Graph for Item No. 5
Any Natural Calamity is the Result of Certain Natural Causes and so to understand the Nature of Calamity, its Antecedents need to be Studied

Pile No.	*Frequency*	*Percentage*	*Cumulative Percentage*
1.	0	0	0
2.	0	0	0
3.	0	0	0
4.	0	0	0
5.	0	0	0
6.	1	3.12	3.12
7.	0	0	3.12
8.	3	9.38	12.50
9.	5	15.63	28.13
10.	9	28.12	56.25
11.	14	43.75	100.00

Median = 9.20, Q=0.8. Item Selected.

Cumulative Frequency Graph for Item No. 6
Raising Theoretical Structres is a Fanciful Romanticism

Pile No.	*Frequency*	*Percentage*	*Cumulative Percentage*
1.	9	32.14	32.14
2.	3	10.72	42.86
3.	1	3.57	46.43
4.	5	17.86	64.29
5.	1	3.57	67.86
6.	1	3.57	71.43
7.	1	3.57	75.00
8.	2	7.14	82.14
9.	4	14.29	96.43
10.	1	3.57	100.00
11.	0	0	100.00

Median = 2.70, Q=3.05. Item Rejected.

Cumulative Frequency Graph for Item No. 7
Critical Faculties are Instrumental in Expanding the Horizons of Knowledge

Pile No.	*Frequency*	*Percentage*	*Cumulative Percentage*
1.	0	0	0
2.	0	0	0
3.	0	0	0
4.	0	0	0
5.	0	0	0
6.	0	0	0
7.	2	6.06	6.06
8.	2	6.06	12.12
9.	3	9.09	21.21
10.	9	27.28	48.49
11.	17	51.51	100.00

Median = 9.5, Q = 0.65. Item Selected.

Cumulative Frequency Graph for Item No. 8
Science is based on Concrete Facts, not on Assumptions

Pile No.	*Frequency*	*Percentage*	*Cumulative Percentage*
1.	1	3.23	3.23
2.	0	0	3.23
3.	2	6.45	9.68
4.	0	0	9.68
5.	1	3.23	12.91
6.	4	12.90	25.81
7.	0	0	25.81
8.	2	6.45	32.26
9.	3	9.68	41.94
10.	5	16.13	58.07
11.	13	41.93	100.00

Median = 9.00, Q=2.25. Item Selected but Eliminated.

Cumulative Frequency Graph for Item No. 9
No Authority is Infallible. So one should not Hesitate to Challenge it.

Pile No.	*Frequency*	*Percentage*	*Cumulative Percentage*
1.	0	0	0
2.	0	0	0
3.	1	3.03	3.03
4.	0	0	3.03
5.	0	0	3.03
6.	1	3.03	6.06
7.	1	3.03	9.09
8.	3	9.09	18.18
9.	1	3.03	21.21
10.	11	33.33	54.54
11.	15	45.46	100.00

Median = 9.4, Q = 0.65. Item Selected.

Cumulative Frequency Graph for Item No. 10
One should not Hesitate to take fast Judgements and Drawing Conclusions

Pile No.	*Frequency*	*Percentage*	*Cumulative Percentage*
1.	8	24.24	24.24
2.	5	15.16	39.40
3.	4	12.12	51.52
4.	0	0	51.52
5.	3	9.09	60.61
6.	4	12.12	72.73
7.	0	0	72.73
8.	4	12.12	84.85
9.	3	9.09	93.94
10.	1	3.03	96.97
11.	1	3.03	100.00

Median = 2.40. Q = 3.10. Item Rejected.

Cumulative Frequency Graph for Item No. 11
Humility does not Signify Confession of Ignorance. Even when one is sure that he knows more than others, he need not give up Humility.

Pile No.	*Frequency*	*Percentage*	*Cumulative Percentage*
1.	0	0	0
2.	1	3.12	3.12
3.	1	3.12	6.24
4.	0	0	0
5.	0	0	0
6.	4	12.50	18.74
7.	2	6.25	24.91
8.	1	3.12	28.11
9.	3	9.39	37.50
10.	8	25.00	62.50
11.	12	37.50	100.00

Median = 9.00. Q = 1.65. Item Selected.

Cumulative Frequency Graph for Item No. 12
One need not Insist for Supporting Proof when great People themselves make Statement of Facts

Pile No.	*Frequency*	*Percentage*	*Cumulative Percentage*
1.	15	45.46	45.46
2.	7	21.21	66.67
3.	2	6.06	72.73
4.	2	6.06	78.79
5.	1	3.03	81.82
6.	2	6.06	87.88
7.	0	0	87.88
8.	0	0	87.88
9.	2	6.06	93.94
10.	1	3.03	96.97
11.	1	3.03	100.00

Median = 0.70, Q = 1.30. Item Selected.

Cumulative Frequency Graph for Item No. 13

As the Old Order changes yielding place to new, one needs to be always ready to understand the New

Pile No.	*Frequency*	*Percentage*	*Cumulative Percentage*
1.	0	0	0
2.	0	0	0
3.	0	0	0
4.	0	0	0
5.	0	0	0
6.	1	3.03	3.03
7.	2	6.06	9.09
8.	3	9.09	18.18
9.	8	24.24	42.42
10.	8	24.24	66.66
11.	11	33.34	100.00

Median = 8.80. Q = 0.95. Item Selected.

Cumulative Frequency Graph for Item No. 14

It is Obvious that often we draw Final Inference First and then Collect Supporting Evidence

Pile No.	*Frequency*	*Percentage*	*Cumulative Percentage*
1.	6	19.35	19.35
2.	0	0	19.35
3.	5	16.13	35.48
4.	1	3.23	38.71
5.	3	9.68	48.39
6.	3	9.67	58.06
7.	6	19.35	77.41
8.	2	6.45	83.86
9.	3	9.68	93.54
10.	1	3.23	96.77
11.	1	3.23	100.00

Median = 4.70, Q = 2.25. Item Selected.

Cumulative Frequency Graph for Item No. 15
I Find it difficult to Resist my curiosity to find out the Whys and Hows of observed Phenomena

Pile No.	*Frequency*	*Percentage*	*Cumulative Percentage*
1.	0	0	0
2.	0	0	0
3.	0	0	0
4.	0	0	0
5.	1	3.12	3.12
6.	0	0	3.12
7.	1	3.12	6.24
8.	3	9.37	15.61
9.	7	21.88	37.49
10.	13	40.63	78.12
11.	7	21.88	100.00

Median = 8.90, Q = 0.75. Item Selected.

Cumulative Frequency Graph for Item No. 16
Quantification which is being Employed these Days is an Untrustworthy Mechanistic Device

Pile No.	*Frequency*	*Percentage*	*Cumulative Percentage*
1.	3	9.38	9.38
2.	7	21.88	31.26
3.	7	21.88	53.14
4.	5	15.62	68.76
5.	4	12.50	81.26
6.	3	9.38	90.64
7.	1	3.12	93.76
8.	1	3.12	96.88
9.	1	3.12	100.00
10.	0	0	100.00
11.	0	·0	100.00

Median = 2.40, Q = 1.40. Item Selected.

Cumulative Frequency Graph for Item No. 17

If I were a Teacher I would Encourage my Students to Question me as closely as they can

Pile No.	*Frequency*	*Percentage*	*Cumulative Percentage*
1.	0	0	0
2.	0	0	0
3.	0	0	0
4.	0	0	0
5.	0	0	0
6.	0	0	0
7.	0	0	0
8.	1	3.03	3.03
9.	2	6.06	9.09
10.	2	6.06	15.15
11.	28	84.85	100.00

Median = 10.00, Q = 0.30. Item Selected.

Cumulative Frequency Graph for Item No. 18

Perfect Precision is a Dream. When observations are made through human medium which is far from perfect. It is unrealistic to aim at perfect precision

Pile No.	*Frequency*	*Percentage*	*Cumulative Percentage*
1.	1	3.23	3.23
2.	0	0	3.23
3.	0	0	3.23
4.	1	3.23	6.46
5.	1	3.23	9.69
6.	4	12.90	22.59
7.	2	6.45	29.04
8.	0	0	29.04
9.	9	29.03	58.07
10.	9	29.03	87.10
11.	4	12.90	100.00

Median = 8.20, Q = 1.65. Item Selected But Eliminated.

Cumulative Frequency Graph for Item No. 19

The Scientists Unending Quest is a part of his Desire to understand the Reality in its Totality

Pile No.	*Frequency*	*Percentage*	*Cumulative Percentage*
1.	1	3.03	3.03
2.	0	0	3.03
3.	1	3.03	6.06
4.	0	0	6.06
5.	0	0	6.06
6.	0	0	6.06
7.	1	3.03	9.09
8.	1	3.03	12.12
9.	6	18.18	30.30
10.	12	36.37	66.67
11.	11	33.33	100.00

Median = 9.00, Q = 0.80. Item Selected.

Cumulative Frequency Graph for Item No. 20

Old Evidences related to a Problem are more Authentic than the Most Recent Ones

Pile No.	*Frequency*	*Percentage*	*Cumulative Percentage*
1.	4	12.50	12.50
2.	7	21.88	34.38
3.	6	18.75	53.13
4.	8	25.00	78.13
5.	2	6.25	84.38
6.	4	12.30	96.88
7.	0	0	96.88
8.	1	3.12	100.00
9.	0	0	100.00
10.	0	0	100.00
11.	0	0	100.00

Median = 2.40, Q = 1.15. Item Selected.

Cumulative Frequency Graph for Item No. 21
As I grow I often find Myself Revising my Opinions and Conclusions

Pile No.	*Frequency*	*Percentage*	*Cumulative Percentage*
1.	0	0	0
2.	0	0	0
3.	0	0	0
4.	1	3.03	3.03
5.	0	0	3.03
6.	0	0	3.03
7.	2	6.06	9.09
8.	6	18.18	27.27
9.	10	30.30	57.57
10.	6	18.18	75.75
11.	8	24.25	100.00

Median = 8.30, Q = 1.00. Item Selected.

Cumulative Frequency Graph for Item No. 22
One need not Rule out the Possibility of a lag between the Drawing of Conclusion and Completeness of Evidence

Pile No.	*Frequency*	*Percentage*	*Cumulative Percentage*
1.	1	3.12	3.12
2.	0	0	3.12
3.	0	0	3.12
4.	0	0	3.12
5.	1	3.12	6.24
6.	2	6.25	12.49
7.	3	9.37	21.86
8.	9	28.13	49.99
9.	9	28.13	78.12
10.	5	15.63	93.75
11.	2	6.25	100.00

Median = 7.50, Q = 0.85. Item Selected But Eliminated.

Cumulative Frequency Graph for Item No. 23

I feel excited when I see a new thing or hear a New Idea

Pile No.	*Frequency*	*Percentage*	*Cumulative Percentage*
1.	0	0	0
2.	0	0	0
3.	1	3.12	3.12
4.	1	3.12	6.24
5.	0	0	6.24
6.	1	3.12	9.36
7.	3	9.38	18.74
8.	4	12.50	31.24
9.	8	25.00	56.24
10.	6	18.76	75.00
11.	8	25.00	100.00

Median = 8.30, Q = 1.25. Item Selected.

Cumulative Frequency Graph for Item No. 24

When the observed Phenomenon is Self Evident, one need not Insist for further Explanation

Pile No.	*Frequency*	*Percentage*	*Cumulative Percentage*
1.	2	6.06	6.06
2.	2	6.06	12.12
3.	7	21.22	33.34
4.	4	12.12	45.46
5.	2	6.06	51.52
6.	4	12.12	63.64
7.	3	9.09	72.73
8.	1	3.03	75.76
9.	2	6.06	81.82
10.	3	9.09	90.91
11.	3	3.03	100.00

Median = 4.20, Q = 2.50. Item Selected.

Cumulative Frequency Graph for Item No. 25
I do not feel Uncomfortable in revising the Approaches I formulate towards People, Things and Ideas

Pile No.	*Frequency*	*Percentage*	*Cumulative Percentage*
1.	0	0	0
2.	0	0	0
3.	1	3.12	3.12
4.	0	0	3.12
5.	0	0	3.12
6.	2	6.24	9.36
7.	1	3.12	12.48
8.	8	25.00	37.48
9.	6	18.76	56.24
10.	8	25.00	81.24
11.	6	18.76	100.00

Median = 8.20, Q = 1.10. Item Selected.

Cumulative Frequency Graph for Item No. 26
In Interpreting Results one has to give Sufficient Attitude to Present Social, Economic and Political Situation

Pile No.	*Frequency*	*Percentage*	*Cumulative Percentage*
1.	3	9.09	9.09
2.	1	3.03	12.12
3.	1	3.03	15.15
4.	1	3.03	18.18
5.	1	3.03	21.21
6.	3	9.09	30.30
7.	3	9.09	39.39
8.	7	21.21	60.60
9.	4	12.12	72.72
10.	4	12.12	84.84
11.	5	15.16	100.00

Median = 7.00, Q = 1.90. Item Selected But Eliminated.

Cumulative Frequency Graph for Item No. 27

It is likely that much of the scientific information we have today will be demonstrated to be in-accurate or inadequate in the future

Pile No.	*Frequency*	*Percentage*	*Cumulative Percentage*
1.	1	3.03	3.03
2.	2	6.06	9.09
3.	5	15.15	24.24
4.	1	3.03	27.27
5.	2	6.06	33.33
6.	4	12.12	45.45
7.	2	6.06	51.51
8.	4	12.12	63.63
9.	6	18.19	81.82
10.	4	12.12	93.94
11.	2	6.06	100.00

Median = 6.30, Q = 2.65. Item Selected But Eliminated.

Cumulative Frequency Graph for Item No. 28

It is the mediocre who Insists on Observation & Recording of facts as the wise can formulate conclusions even without the crude data.

Pile No.	*Frequency*	*Percentage*	*Cumulative Percentage*
1.	11	33.33	33.33
2.	4	12.12	45.45
3.	4	12.12	57.57
4.	5	15.15	72.72
5.	2	6.06	78.78
6.	5	15.15	93.93
7.	1	3.03	96.96
8.	0	0	96.96
9.	0	0	96.96
10.	1	3.03	100.00
11.	0	0	100.00

Median = 1.90, Q = 1.75. Item Selected.

Cumulative Frequency Graph for Item No. 29

I do not hesitate to receive any evidence, writen or spoken which opposes my strongest personal opinion

Pile No.	*Frequency*	*Percentage*	*Cumulative Percentage*
1.	0	0	0
2.	0	0	0
3.	0	0	0
4.	0	0	0
5.	1	3.03	3.03
6.	0	0	3.03
7.	0	0	3.03
8.	5	15.15	18.18
9.	9	27.27	45.45
10.	7	21.21	66.66
11.	11	33.33	100.00

Median = 8.80, Q = 1.00. Item Selected.

Cumulative Frequency Graph for Item No. 30

When the findings or Theories of Science conflict with religious belief, it is better to accept the religious belief.

Pile No.	*Frequency*	*Percentage*	*Cumulative Percentage*
1.	13	39.39	39.39
2.	6	18.18	57.57
3.	5	15.15	72.72
4.	3	9.09	81.81
5.	1	3.03	84.84
6.	4	12.13	96.97
7.	0	0	96.97
8.	0	0	96.97
9.	1	3.03	100.00
10.	0	0	100.00
11.	0	0	100.00

Median = 1.10, Q = 1.20. Item Selected.

Cumulative Frequency Graph for Item No. 31

I hate all Superstitions which Mosquarade as Traditions in our country

Pile No.	*Frequency*	*Percentage*	*Cumulative Percentage*
1.	4	12.12	12.12
2.	3	9.09	21.21
3.	1	3.03	24.24
4.	0	0	24.24
5.	3	9.09	33.33
6.	4	12.12	45.45
7.	0	0	45.45
8.	2	6.06	51.51
9.	4	12.12	63.63
10.	7	21.21	84.84
11.	5	15.16	100.00

Median = 7.50, Q = 2.70. Item Selected.

Cumulative Frequency Graph for Item No. 32

There are many things in the world which transcend human rationality and so must be accepted

Pile No.	*Frequency*	*Percentage*	*Cumulative Percentage*
1.	4	12.50	12.50
2.	2	6.24	18.74
3.	3	9.38	28.12
4.	6	18.77	46.89
5.	1	3.12	50.01
6.	6	18.77	68.78
7.	2	6.24	75.02
8.	4	12.50	87.52
9.	2	6.24	93.76
10.	1	3.12	96.88
11.	1	3.12	100.00

Median = 4.50, Q = 2.20. Item Selected.

Cumulative Frequency Graph for Item No. 33

I prefer to accept facts which are rationally explained

Pile No.	*Frequency*	*Percentage*	*Cumulative Percentage*
1.	0	0	0
2.	0	0	0
3.	0	0	0
4.	0	0	0
5.	0	0	0
6.	0	0	0
7.	3	9.09	9.09
8.	6	18.18	27.27
9.	3	9.09	36.36
10.	8	24.24	60.60
11.	13	39.39	100.00

Median = 9.10, Q = 1.25. Item Selected.

Cumulative Frequency Graph for Item No. 34

The traditional views have survived over the centuries because of their validity and so they must be accepted

Pile No.	*Frequency*	*Percentage*	*Cumulative Percentage*
1.	9	27.28	27.28
2.	4	12.12	39.40
3.	3	9.09	48.49
4.	1	3.03	51.52
5.	3	9.09	60.61
6.	7	21.21	81.82
7.	2	6.06	87.88
8.	1	3.03	90.91
9.	2	6.06	96.97
10.	1	3.03	100.00
11.	0	0	100.00

Median = 3.20, Q = 2.25. Item Selected.

Cumulative Frequency Graph for Item No. 35

A good scientist always demonstrates concern for observation and recording of facts

Pile No.	*Frequency*	*Percentage*	*Cumulative Percentage*
1.	0	0	0
2.	0	0	0
3.	0	0	0
4.	0	0	0
5.	1	3.03	3.03
6.	0	0	3.03
7.	1	3.03	6.06
8.	2	6.06	12.12
9.	5	15.15	27.27
10.	9	27.27	54.54
11.	15	45.45	100.00

Median = 9.00, Q = 0.80. Item Selected.

Cumulative Frequency Graph for Item No. 36

Gandhiji has asserted that the Bihar earthquake was a punishment inflicted by the nature on Biharis for the sin of untouchability

Pile No.	*Frequency*	*Percentage*	*Cumulative Percentage*
1.	19	57.57	57.57
2.	3	9.09	66.66
3.	2	6.06	72.72
4.	2	6.06	78.78
5.	1	3.03	81.81
6.	6	18.18	100.00
7.	0	0	100.00
8.	0	0	100.00
9.	0	0	100.00
10.	0	0	100.00
11.	0	0	100.00

Median = 0.50, Q = 1.35. Item Selected.

Cumulative Frequency Graph for Item No. 37

The vagaries of present social, economic and political situation should not modify the interpretation of results

Pile No.	*Frequency*	*Percentage*	*Cumulative Percentage*
1.	1	3.03	3.03
2.	0	0	3.03
3.	3	9.09	12.12
4.	3	9.09	21.21
5.	0	0	21.21
6.	6	18.18	39.39
7.	1	3.03	42.42
8.	6	18.18	60.60
9.	4	12.12	72.72
10.	3	9.09	81.81
11.	6	18.18	100.00

Cumulative Frequency Graph for Item No. 38

Being always critical is not a virtue

Pile No.	*Frequency*	*Percentage*	*Cumulative Percentage*
1.	2	6.16	6.06
2.	4	12.12	18.18
3.	2	6.06	24.24
4.	2	6.06	30.30
5.	1	3.03	33.33
6.	7	21.21	54.54
7.	4	12.12	66.66
8.	2	6.06	72.72
9.	4	12.12	84.84
10.	1	3.03	87.87
11.	4	12.12	100.00

Median = 5.30, Q = 250. Item Selected.

Cumulative Frequency Graph for Item No. 39
Seeking further information about a novel situation is considered as a sound approach

Pile No.	*Frequency*	*Percentage*	*Cumulative Percentage*
1.	0	0	0
2.	0	0	0
3.	0	0	0
4.	1	3.03	3.03
5.	0	0	3.03
6.	2	6.06	9.09
7.	1	3.03	12.12
8.	7	21.21	33.33
9.	8	24.24	57.57
10.	5	15.15	72.72
11.	9	27.27	100.00

Median = 8.20, Q = 1.25. Item Selected.

Cumulative Frequency Graph for Item No. 40
Some, because of their Expertise and Mastery over a field of human activity are considered to be the authorities. It woud be unwise to challenge them

Pile No.	*Frequency*	*Percentage*	*Cumulative Percentage*
1.	10	30.30	30.30
2.	6	18.18	48.48
3.	4	12.12	60.69
4.	5	15.15	75.75
5.	4	12.12	87.87
6.	2	6.06	93.93
7.	0	0	93.93
8.	0	0	93.93
9.	0	0	93.93
10.	2	6.06	100.00
11.	0	0	100.00

Median = 1.70, Q = 1.55. Item Selected.

Cumulative Frequency Graph for Item No. 41

Conclusions could be drawn only when the evidence is sensible and complete

Pile No.	*Frequency*	*Percentage*	*Cumulative Percentage*
1.	0	0	0
2.	0	0	0
3.	0	0	0
4.	1	3.03	3.03
5.	1	3.03	6.06
6.	1	3.03	9.09
7.	4	12.12	21.21
8.	4	12.12	33.33
9.	8	24.24	57.57
10.	4	12.12	69.69
11.	10	30.30	100.00

Median = 8.20, Q = 1.40. Item Selected But Eliminated.

Cumulative Frequency Graph for Item No. 42

It is hypocracy to demonstrate humility when one is sure that he knows more than others

Pile No.	*Frequency*	*Percentage*	*Cumulative Percentage*
1.	5	15.63	15.63
2.	3	9.38	25.01
3.	1	3.12	28.13
4.	6	18.75	46.88
5.	0	0	46.88
6.	10	31.25	78.13
7.	2	6.25	84.38
8.	0	0	84.38
9.	3	9.38	93.96
10.	2	6.24	93.96
11.	0	0	100.00

Median = 4.70, Q = 1.95. Item Selected.

Cumulative Frequency Graph for Item No. 43
When a gamut of authoritative evidence relating to a problem are furnished I prefer to select the most recent one

Pile No.	*Frequency*	*Percentage*	*Cumulative Percentage*
1.	2	6.45	6.45
2.	1	3.23	9.68
3.	0	0	9.68
4.	4	12.90	22.58
5.	4	12.90	35.48
6.	6	19.36	54.84
7.	3	9.68	64.52
8.	2	6.45	70.97
9.	4	12.90	83.87
10.	3	9.68	93.55
11.	2	6.45	100.00

Median = 5.20, Q = 2.05. Item Selected.

Cumulative Frequency Graph for Item No. 44
All the possible patterns are already conceived by the great theoreticians I adore. Nothing new which has not already been theorised can emerge which marits renewed understanding.

Pile No.	*Frequency*	*Percentage*	*Cumulative Percentage*
1.	16	48.48	48.48
2.	4	12.12	60.60
3.	4	12.12	72.72
4.	2	6.06	78.78
5.	2	6.06	84.84
6.	2	6.06	90.90
7.	1	3.03	93.93
8.	1	3.03	96.96
9.	1	3.03	100.00
10.	0	0	100.00
11.	0	0	100.00

Median = 0.60, Q = 1.35. Item Selected

Cumulative Frequency Graph for Item No. 45

Accuracy and precision are the cornerstones of Science

Pile No.	*Frequency*	*Percentage*	*Cumulative Percentage*
1.	0	0	0
2.	1	3.12	3.12
3.	0	0	3.12
4.	0	0	3.12
5.	1	3.12	6.24
6.	2	6.25	12.49
7.	1	3.12	15.61
8.	5	15.63	31.24
9.	1	3.12	34.36
10.	6	18.75	53.11
11.	15	46.88	100.00

Median = 9.30, Q = 1.45. Item Selected But Eliminated.

Cumulative Frequency Graph for Item No. 46

We must be concerned with the phenomenon we confront with. It is sheer adventurism to seek to unearth the antecedents of each observed phenomenon.

Pile No.	*Frequency*	*Percentage*	*Cumulative Percentage*
1.	4	12.12	12.12
2.	4	12.12	24.24
3.	4	12.12	36.36
4.	3	9.09	45.45
5.	2	6.16	51.51
6.	6	18.18	69.69
7.	6	18.18	87.87
8.	0	0	87.87
9.	3	9.09	96.96
10.	1	3.03	100.00
11.	0	0	100.00

Median = 4.20, Q = 2.10. Item Selected.

Cumulative Frequency Graph for Item No. 47
No meaningful conclusions can be drawn unless the evidence is quantified

Pile No.	*Frequency*	*Percentage*	*Cumulative Percentage*
1.	1	3.03	3.03
2.	0	0	3.03
3.	3	9.09	12.12
4.	3	9.09	21.21
5.	2	6.06	27.27
6.	3	9.09	36.36
7.	2	6.06	42.42
8.	4	12.12	54.54
9.	3	9.09	63.63
10.	6	18.18	81.81
11.	6	18.18	100.00

Median = 7.10, Q = 2.45. Item Selected.

Cumulative Frequency Graph for Item No. 48
When candid and lucid narrations can make the students properly understand the things to be taught. It is (naive) pedantic to resort to questioning method.

Pile No.	*Frequency*	*Percentage*	*Cumulative Percentage*
1.	5	16.13	16.13
2.	3	9.68	25.81
3.	4	12.90	38.71
4.	4	12.90	51.61
5.	5	16.13	67.74
6.	4	12.90	80.64
7.	3	9.98	90.32
8.	1	3.23	93.55
9.	0	0	93.55
10.	1	3.23	96.78
11.	1	3.23	100.00

Median = 3.30, Q = 1.85. Item Selected.

Cumulative Frequency Graph for Item No. 49
I do not feel like drawing final inferences before convincing evidence is collected

Pile No.	*Frequency*	*Percentage*	*Cumulative Percentage*
1.	0	0	0
2.	0	0	0
3.	0	0	0
4.	2	6.06	6.06
5.	0	0	6.06
6.	1	3.03	9.09
7.	1	3.03	12.12
8.	6	18.18	30.30
9.	4	12.12	42.42
10.	9	27.27	69.69
11.	10	30.30	100.00

Median = 8.70, Q = 1.20. Item Selected.

Cumulative Frequency Graph for Item No. 50
Man has already the command over the totality of knowledge. Still greater efforts in the directions of knowing more are uncalled for.

Pile No.	*Frequency*	*Percentage*	*Cumulative Percentage*
1.	20	60.60	60.60
2.	5	15.15	75.75
3.	4	12.12	87.87
4.	0	0	87.87
5.	1	3.03	90.90
6.	1	3.03	93.93
7.	0	0	93.93
8.	1	3.03	96.96
9.	0	0	96.96
10.	0	0	96.96
11.	1	3.03	100.00

Median = 0.40, Q = 0.65. Item Selected.

Cumulative Frequency Graph for Item No. 51

One need not accept the statements emanating from great men as facts unless they are supported by convincing proof

Pile No.	*Frequency*	*Percentage*	*Cumulative Percentage*
1.	1	3.03	3.03
2.	1	3.03	6.06
3.	0	0	6.06
4.	1	3.03	9.09
5.	1	3.03	12.12
6.	0	0	12.12
7.	1	3.03	15.15
8.	5	15.15	30.30
9.	2	6.06	36.36
10.	10	30.30	66.66
11.	11	33.33	100.00

Median = 9.00, Q = 1.30. Item Selected.

Cumulative Frequency Graph for Item No. 52

As I am not in the habit of entertaining wrong opinions and defective conclusions. I need for change them.

Pile No.	*Frequency*	*Percentage*	*Cumulative Percentage*
1.	8	24.24	24.24
2.	4	12.12	36.36
3.	9	27.27	63.63
4.	4	12.12	75.75
5.	1	3.03	78.78
6.	1	3.03	81.81
7.	1	3.03	84.84
8.	1	3.03	87.87
9.	2	6.06	93.93
10.	0	0	93.93
11.	2	6.06	100.00

Median = 2.00, Q = 1.40. Item Selected.

Cumulative Frequency Graph for Item No. 53

I feel hesitant to take snap judgements when the evidence does not seem to warrant such fast judgements

Pile No.	*Frequency*	*Percentage*	*Cumulative Percentage*
1.	0	0	0
2.	0	0	0
3.	0	0	0
4.	0	0	0
5.	0	0	0
6.	4	12.12	12.12
7.	1	3.03	15.15
8.	7	21.21	36.36
9.	5	15.15	51.51
10.	7	21.21	72.72
11.	9	27.27	100.00

Median = 8.50, Q = 1.30. Item Selected But Eliminated.

Cumulative Frequency Graph for Item No. 54

I do not feel comfortable when I comfront things and ideas which I did not come across earlier.

Pile No.	*Frequency*	*Percentage*	*Cumulative Percentage*
1.	6	18.75	18.75
2.	4	12.50	31.25
3.	2	6.25	37.50
4.	2	6.25	43.75
5.	1	3.12	46.87
6.	5	15.63	62.50
7.	4	12.50	75.00
8.	3	9.38	84.38
9.	3	9.38	93.76
10.	0	0	93.76
11.	2	6.25	100.00

Median = 4.70, Q = 2.75. Item Selected.

Cumulative Frequency Graph for Item No. 55

Scientific Theory is based on certain assumptions. The moment the assumptions are proved wrong, the entire citadel of science collapses

Pile No.	*Frequency*	*Percentage*	*Cumulative Percentage*
1.	3	9.09	9.09
2.	3	9.09	18.18
3.	3	9.09	27.27
4.	4	12.12	39.39
5.	3	9.09	48.48
6.	3	9.09	57.57
7.	2	6.06	63.63
8.	3	9.09	72.72
9.	4	12.12	84.84
10.	1	3.03	87.87
11.	4	12.12	100.00

Median = 4.70, Q = 2.75. Item Selected But Eliminated.

Cumulative Frequency Graph for Item No. 56

When I am convinced that the approach I have adopted is perfect why should I adopt multiple alternative approaches to people, things and ideas

Pile No.	*Frequency*	*Percentage*	*Cumulative Percentage*
1.	4	12.50	12.50
2.	5	15.63	28.13
3.	6	18.75	46.88
4.	4	12.50	59.38
5.	5	12.50	71.88
6.	4	12.50	84.38
7.	0	0	84.38
8.	1	3.12	87.50
9.	0	0	87.50
10.	4	12.50	100.00
11.	0	0	100.00

Median = 2.70, Q = 1.70. Item Selected.

Cumulative Frequency Graph for Item No. 57
The theoretical structures which are raised so strenuously need to be respected

Pile No.	*Frequency*	*Percentage*	*Cumulative Percentage*
1.	2	6.24	6.24
2.	3	9.37	15.61
3.	3	9.38	24.99
4.	2	6.25	31.24
5.	2	6.25	37.49
6.	3	9.38	46.87
7.	6	18.75	65.62
8.	2	6.25	71.87
9.	3	9.38	81.25
10.	4	12.50	93.75
11.	2	6.25	100.00

Median = 5.70, Q = 2.70. Item Selected But Eliminated.

Cumulative Frequency Graph for Item No. 58
As the Truth is unchanging the ideas about the Truth should remain unchanged

Pile No.	*Frequency*	*Percentage*	*Cumulative Percentage*
1.	6	18.75	18.75
2.	6	18.75	37.50
3.	5	15.63	53.13
4.	3	9.38	62.51
5.	3	9.38	71.89
6.	4	12.50	84.39
7.	2	6.25	90.64
8.	0	0	90.64
9.	1	3.12	93.76
10.	0	0	93.76
11.	2	6.25	100.00

Median = 2.30, Q = 1.80. Item Selected.

Cumulative Frequency Graph for Item No. 59

There are always alternative competing probabilities which deserve to be accepted

Pile No.	*Frequency*	*Percentage*	*Cumulative Percentage*
1	0	0	0
2.	0	0	0
3.	0	0	0
4.	1	3.33	3.33
5.	0	0	3.33
6.	2	6.67	10.00
7.	2	6.67	16.67
8.	6	20.00	36.67
9.	6	20.00	56.67
10.	6	20.00	76.67
11.	7	23.33	100.00

Median = 8.10, Q = 1.25. Item Selected.

Cumulative Frequency Graph for Item No. 60

Obviously I donot feel very comfortable when I listen to observe or read evidence contradicting my strong personal opinions

Pile No.	*Frequency*	*Percentage*	*Cumulative Percentage*
1	5	16.67	16.67
2.	4	13.33	30.00
3.	5	16.67	46.67
4.	1	3.33	50.00
5.	4	13.33	63.33
6.	2	6.67	70.00
7.	2	6.67	76.67
8.	2	6.67	83.34
9.	0	0	83.34
10.	5	16.67	100.00
11.	0	0	100.00

Median = 3.50, Q = 2.60. Item Selected.

Cumulative Frequency Graph for Item No. 61

When evidence warrants a new generalization, one should not hesitate to shed the old and accept the new generalization

Pile No.	*Frequency*	*Percentage*	*Cumulative Percentage*
1	0	0	0
2.	0	0	0
3.	0	0	0
4.	1	3.12	3.12
5.	0	0	3.12
6.	1	3.12	6.24
7.	2	6.25	12.49
8.	2	6.25	18.74
9.	3	9.38	28.12
10.	7	21.88	50.00
11.	16	50.00	100.00

Median = 9.50, Q = 0.90. Item Selected But Eliminated.

Cumulative Frequency Graph for Item No. 62

Even the so called superstition has deep roots in reality which we often fail to take into congnisance

Pile No.	*Frequency*	*Percentage*	*Cumulative Percentage*
1	4	12.90	12.90
2.	1	3.23	16.13
3.	3	9.63	25.81
4.	1	3.23	29.04
5.	3	9.68	38.72
6.	10	32.26	70.98
7.	2	6.45	77.53
8.	2	6.45	83.88
9.	4	12.90	96.78
10.	1	3.23	100.00
11.	0	0	100.00

Median = 4.80, Q = 1.83. Item Selected.

Bibliography

Abder, Pamela Fraser and Shrigley, Robert C. "A status study of the Scientific Attitudes of Elementary School Teachers in Trinidad and Tobago", *Science Education* 64 (5): 637-644 (1980), John Wiley & Sons,Inc.

Aiken, Lewis R.Jr and Aiken, Dorothy R. "Recent Research on Attitudes concerning Science", *Science Education.* No. 4 (1969).

Anderson, Ronald D., et al. *Developing Children's Thinking through Science,* Prentice-Hall, Inc., New Jersey (1970).

Baumel, H.B. and Berger, J. Joel, "An attempt to measure Scientific Attitude", *Science Education* (1965) 49: 267-269.

Billeh. Victor Y. and Zakhariades, George A. "The Development and Application of a Scale for Measuring Scientific Attitudes", *Science Education* 59(2): 155-165 (1975) John Wiley & Sons., Inc.

Burnett, R..W. *The Science Teacher and his Objectives,* Teachers College Record. 1944, 45.

Carin.Arthur A. and Sund. Robert B., *Teaching Science through Discovery* Second Edn., Charles E Merill Publishing Company, Columbus, Ohio, 1970.

Diederich, Paul B. "Components of Scientific Attitudes", *The Science Teacher,* 34: 23-24, (1967).

Educational Policies Commission and the spirit of Science, Washington D.C: National Education Association, 1966.

Edwards A.L. *Techniques of Attitude Scale Construction,* New York: Appleton-Century-Crofts, Inc. 1957.

Garrett, Henry E. *Statistics in Psychology and Education,* Vakils, Feffer and Simons Ltd. Vakils House, Bombay.

Gauld, Colin. "The Scientific Attitude and Science Education: A Critical Reappraisal", *Science Education,* 66 (11:109-121) (1982). John Wiley & Sons., Inc.

Ghosh.B.N. *Lectures on Scientific Method,* Sterling Publishers Private Limited, New Delhi, 1986.

Haney, R.E. "The Development of Scientific Attitudes", *Science Teacher,* 31: 33-35 (1964).

Heiss, Elwood D. *"Modern Science Teaching",* Ellsworths, Obourn and Charles W. Hoffman; Macmillan Company, New York, 1961.

"International Encyclopedia of Educational Research and Studies", 1985, Vol. 8S, pp. 4450-4453.

Kozlow, M.J. and Nay, M.A. "An Approach to Measuring Scientific Attitudes", *Science Education* 60 (2): 147-172 (1976) . John Wiley & Sons., Inc.

Ladd. Everett Carll, Jr. and Lipset, Seymour Martin. *"The Divided Academy-Professors and Politics",* McGraw-Hill Book Company, New York, 1975.

National Society for the Study of Education, *"Science and Secondary Schools,"* in Science Education in American Schools. The Forty-sixth year book Part-I Chicago: The University of Chicago Press, 1947, pp. 147-149.

Noll, V.H. *"The Habit of Scientific Thinking",* Teachers College Record, 1933 a, 35, 1-9.

———"Measuring the Scientific Attitude", *The Journal of Abnormal and Social Psychology*, 30: 145-154 (1935).

Oppenheim, A.N. "*Questionnaire Design and Attitude Measurement*," New York, Basic Books, Inc. 1956.

Rethinking Science Education, Fifty-ninth year book of the National Society for the Study of Education, Chicago, Illinois. University of Chicago Press, 1960, p. 46.

Shah, A.B., *Scientific Method*, Allied Publishers Private Limited, Bombay, 1964.

Third Survey of Education Research, CASE, Baroda, 1987.

Van Dalen, D. B. "Relationship of Fact and Theory in Research," *Educational Administration and Supervision*, 45 (1959): 271-274.

Index

❑❑❑